Unveiling Bee Decline's Causes

Chris

CONTENTS

CHAPTER 1. INTRODUCTION

Ecosystem services are defined as services that contribute directly and indirectly to human well-being (Braat & de Groot, 2012). Pollination, as well as pest control and seed dispersal, are examples of ecosystem services produced by mobile organisms that forage within or between habitats (Lundberg, 2003). Flowering plant species pollinated by animals experience an increase in both the size and quality of crops, with bees proving to be the most important pollinators (Roubik, 2001). Therefore, it's possible to ensure the conservation of habitats through the conservation of bee populations, 'for if the bees disappear, reproduction of major elements of the flora may be severely limited' (Michener, 2007), consequently, herbivorous or seed-eating insects, birds and small mammals will be deprived of food and/or host plants, resulting in general loss of species diversity.

Crop pollination by animals is for instance a highly valuable ecosystem service. Almost 90% of angiosperms require pollinators for their reproduction. It has been found that 87 of the 115 leading global crops benefit significantly from pollinators, representing 35% of the global food production (Klein *et al.*, 2007). Not just crop quality but also shelf life and commercial value are improved by bee pollination (Klatt *et al.*, 2014). It has been found that crops are more effectively pollinated when visited by wild insects, doubling fruit sets when compared to the same amount of honey bee visitation (Garibaldi *et al.*, 2013). Additionally, a high proportion (94%) of the wild flowering-plant communities in the tropics are dependent on wild pollinators (Ollerton *et al.*, 2011).

The decline in abundance and diversity of wild bees and abundance of honeybees due to changes in land use and agriculture has been well documented (Biesmeijer, 2006; Olroyd, 2007; Potts, 2010). In order to react to pollinator limitations, farmers have introduced managed flower visitors (as e.g. *Apis mellifera*, honeybees) within crop fields. Despite their effectiveness as pollinator, reliance on a single managed pollinator species, whose numbers are falling sharply, has proved to be risky (Neumann & Carreck, 2010).

Distinctively, worldwide there are around 20,000 native wild bee species (Michener, 2007) and most contribute to the pollination of crops (e.g. Klein, 2003; Kremen, 2002; Klein, 2007; Greenleaf, 2006ab). These native bees biologically complement the honey bee service by enhancing the efficacy of pollination (Greenleaf, 2006b), consequently, conservation and restoration of the habitat used by these organisms offer alternative for reducing dependence on managed honey bees (Kremen, 2002; Rader *et al.*, 2009; Rader *et al.*, 2012). In addition, they also provide an alternative for the pollination scarcity, being therefore economically important (Winfree *et al.*, 2009). It is clear that, when compensation is possible, it is dependent on the functional diversity of wild bee community as well as distribution of habitat and resources of the surrounding landscape. Subsequently, bee functional diversity increases the incidence of fruit and seed set in orchards (Martins *et al.*, 2015).

Rapid human population growth generates an increase in global food demand (Godfray *et al.*, 2010) posing major challenges for increasing crop yield. Concomitantly, natural habitat cover decreased (Hansen *et al.*, 2010) and global stock of pollinators was altered reducing the capacity of ecosystem services to support human activity (Potts, 2010; Garibaldi *et al.*, 2011a).

Although developing countries, like Angola, are still relatively rich in natural habitat, the economic growth is pressing the replacement of natural areas by agricultural landscapes, making them susceptible to pollination problems (Gemmil-Herren & Ochieng, 2008). As suggested by Keating *et al.* (2010), food security and sovereignty should increase directly in areas where hunger exists, based on robust and eco-efficient solutions for agriculture intensification, incorporating both natural biodiversity patterns and processes. Considering the above mentioned, the increase in production should be accompanied by well-informed regional and targeted solutions (Phalan *et al.*, 2011), in order to link agricultural intensification with biodiversity conservation and hunger reduction.

Given the broad ecological and agricultural importance of bees, a profile of the bee fauna of a region would serve as a foundation for subsequent studies or more specific studies, including those involving ecosystem services provided by bees in formal and informal agricultural practices in Angola. As for most insect groups in Angola, there is little published information, and more intensive collecting, taxonomy and ecological work is required if we are to benefit maximally from the pollinator services supplied by the native bee fauna.

1.1. GLOBAL BEE DIVERSITY

Michener (1979) indicated that bee diversity and abundance was higher in warm temperate, xeric regions of the world, apparently an exception to the generalised latitudinal diversity gradient, where organisms and habitat types increase in richness toward the tropics (Hillebrand, 2004). His assessment was based on the taxonomic literature and he acknowledged that the result could be in part due to "factors other than genuine faunal differences" (Michener 1979), with factors such as incomplete taxonomy, and varying intensity of study of local bee faunas biasing the true patterns of regional biodiversity. According to his findings, particularly rich bee faunas are the Mediterranean basin eastward to central Asia, as well as the Madrean region of North America (Mediterranean climatic zones of California, Californian deserts northwest of the Sonoran region and northern Mexico – Chihuahua desert). He also acknowledged the diverse bee faunas of central Chile and Argentina, Australia and western parts of southern Africa. In spite of data limitation, Michener (1979) still considered the bee fauna of the African tropics to be richer than that of the Oriental tropics.

A phylogeny and biogeographical assessment of the Osmiini in particular, indicates a Palaearctic origin for the osmiine bees, showing more diversity in the Old World. The barriers formed by deserts and Afrotropical zones may be reflected in the low dispersal between Palaearctic and sub-Saharan Africa (Praz *et al.*, 2008). For the colletid bees, the expansion of Australia and South America arid biomes is coincident with higher diversification rates of these bees (Almeida *et al.*, 2012). In South America this is because neotropical lineages associated with subtropical and temperate climates were separated on numerous occasions east and west of the Andes (Almeida et al., 2018).

There are exceptions to the general pattern found by Michener with certain bee taxa - Nomiinae, Ctenoplectrini, Xylocopini, Tetrapediini, Centridini, Ericrocidini, Rhathymini,

Meliponini, Apini and Euglossini - having their greatest diversity in the tropics when compared to xeric warm temperate areas. These exceptions could be related to the fact that most of these groups nest in wood, cavities, termite nests or even line their cells with resin or wax being therefore less subject to damages provoked by moisture and fungi (Michener, 1979). The Euglossini tribe, for instance, are distributed throughout the Neotropical region, where they are prominent pollinators (Ramirez *et al.*, 2010).

For four decades since Michener's assessment, accumulating evidence for declines and losses of pollinators have been reported, raising concern over the sustainability of pollination services (Jennersten, 1988; Kearns *et al.*, 1998; Cunningham, 2000; Biesmeijer et al, 2006; Kluser & Peduzzi, 2007; Tylianakis, 2013; Kerr et al, 2015). The evidence applies to both wild and domesticated pollinators, as well as to the plants relying upon them (Steffan-Dewenter *et al.*, 2005; Potts *et al.*, 2010). These declines appear to be driven by human activities (Brown & Paxton, 2009). Land-use intensification, climate change and the spread of alien species and diseases are the main anthropogenic pressures (Vanbergen & IPI, 2013). Evidence from non-organic farms indicate a reduced diversity and abundance of native bees (Kremen *et al.*, 2002), while natural habitat areas in a given landscape increase stability of pollination services (Kremen *et al.*, 2004; Koh *et al.*, 2016). Additionally, the plant-pollinator interaction network has been degrading in the last century with consequent decline in pollination services provided and is estimated to be less resilient to future disturbances (Burkle *et al.*, 2013).

Given the importance of bees, including the non-Apis (Garibaldi *et al.*, 2013) as major pollinators of crops and wild plants (Klein *et al.*, 2007; Ollerton *et al.*, 2011), several initiatives arose encouraging long-term monitoring of their diversity and abundance, which generated an upsurgence in bee conservation studies (API *et al.*, 2003; Environment Ministry, 1999). Consequently, the bee faunas of various regions were subject to more intensive research interest. An overall picture for bee diversity worldwide can be found in Figures 1.1 and 1.2, with countries containing the highest numbers in bee species covered in different green shades.

Over the last few decades, Australia (Figure 1.1) increased their species list to 1,647, with approximately 100 of the bee species being or suspected to be oligolectic (Batley &

Hogendoorn, 2009). Studies suggest that Andrenidae and Melittidae, groups widely distributed globally, are absent from the Australian bee fauna, while the family Stenotritidae and subfamily Euryglossinae are endemic to that region, which also supports disproportionately high number of species of Colletidae. In spite of this resurgence in bee taxonomy in Australia, there are still a large number of undescribed species (Batley & Hogendoorn, 2009).

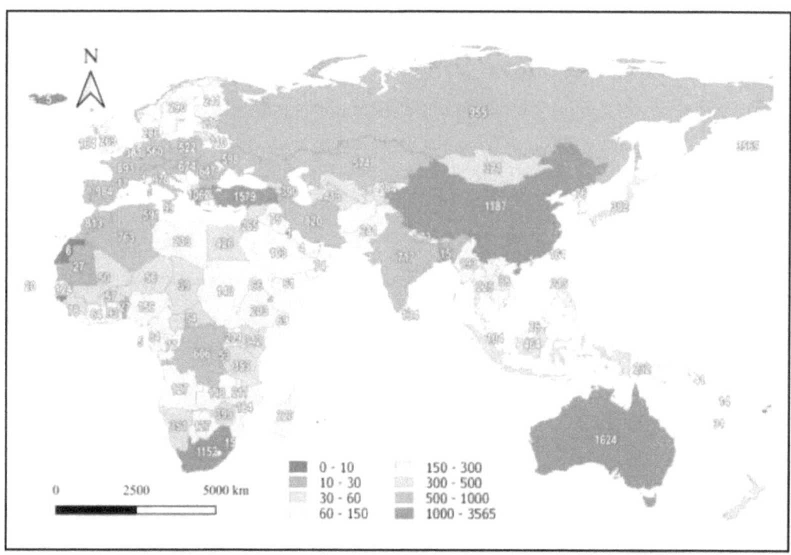

Figure 1.1 - Map representing the bee species richness in the African, Asian, European and Oceania continents. Legend indicates highest bee richness in different shades of green. Adapted from Discover Life (http://www.discoverlife.org/nh/cl/counts/Apoidea_species.html)

In the Neotropics, Central and South America (Figure 1.2), studies in bee diversity and abundance have also recently been intensified. Altogether, the region holds 5,000 described species, it is estimated to have more than 15,000 (Freitas *et al.*, 2009). The tribe Meliponini is extraordinarily rich in the region, inhabiting most of the tropical Americas, from Mexico to Argentina, and to Uruguay, spanning a wide altitudinal range (sea level to 4000m a.s.l.), with 33 valid genera and 391 species native to the region (Freitas *et al.*, 2009). The bee fauna in Mexico represents one of the richest in the world and is estimated to have 1,800 species, in 144 genera and 8 families (four of the genera and eight subgenera being regional endemics). However, again the number of species may be underestimated since some parts of the country have barely been surveyed. Brazil has 1,678 described bee species but more than 50% of the country's territory, including the biomes of the Amazon rainforest and the tropical wetland of Pantanal remain relatively unsampled (Freitas *et al.*, 2009).

In the West-Palaearctic region, the profile of the European bee fauna (Figure 1.1) is the most robust globally, despite significant gaps in coverage, and sufficient to enable accurate analyses on diversity and distributional trends. A revision of the region revealed 2,065 described species, distributed along a North-South positive linear gradient, with Finland having the lowest species richness (241 species) and Spain recording the highest (1,017 species), albeit the latter likely to be an underestimation (Patiny *et al.*, 2009). The knowledge on the bee fauna of the eastern part of the West-Palaearctic is fragmentary but it is assumed that Turkey, Iran and former Mesopotamia have the richest bee faunas (Figure 1.1). The limited number of studies done in the southern West-Palaeartic (Figure 1.1), Saharan area, suggest that the area between western Egypt and southeastern Tunisia has the lowest number of species, with Maghreb and the Nile containing the highest levels of species richness, and with Morocco being a notable hotspot for apifauna richness (Patiny *et al.*, 2009).

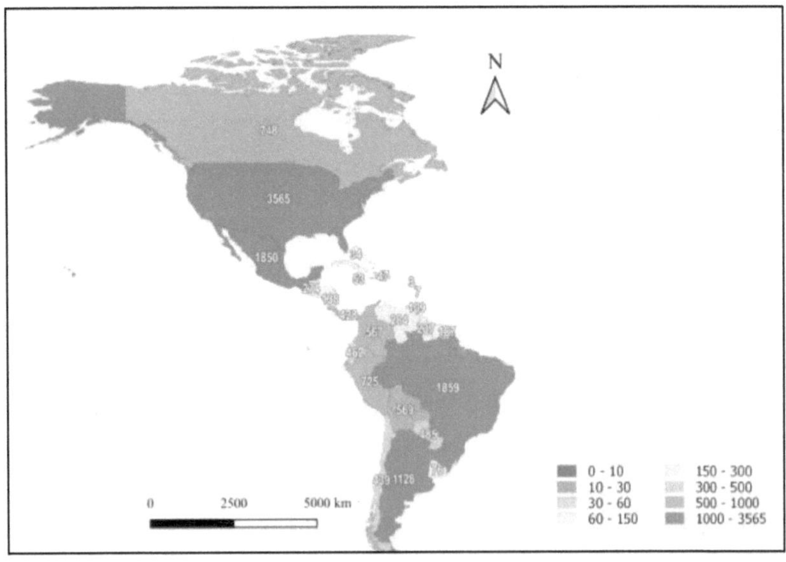

Figure 1.2 - Map representing the bee species richness in the American continent. Legend indicates highest bee richness in different shades of green. Adapted from Discover Life (http://www.discoverlife.org/nh/cl/counts/Apoidea_species.html).

As with many other parts of the world, the apifauna of African countries (Figure 1.1) is largely poorly-known. Where surveys exist, they are not evenly distributed geographically, tending to concentrate in particular regions, leaving a larger part of the continent unsurveyed. Data have been gathered for bee diversity in Southern (Kuhlmann, 2009) and East Africa, in natural habitats (Chiawo *et al.*, 2017), as well as in areas of cultivated and wild crops (Karanja *et al.*, 2010; Morimoto et al., 2004; Njoroge *et al.*, 2004), but syntheses on patterns of diversity and abundance in major biomes is still scarce. The bee diversity in sub-Saharan Africa is considered intermediate when compared to other areas globally, since 6 of the 7 bee families and 129 genera recognized by Michener (2007) occur in the region (Eardley, 2002). The number of described species is approximately 2,600 (Eardley *et al.*, 2009).

1.2.   BEE DIVERSITY IN AFRICA

Generally, in Africa, socio-economic and cultural circumstances are immeasurably different from those in Europe or North America with limitations imposed on scientific endeavor. Data on African bee diversity are as scarce as are specialists for the different African regions (Eardley *et al.*, 2009).

Knowledge of the North African apifauna is fragmentary and mostly restricted to focus studies on a few species. Genetic studies have been done on honeybee (*Apis mellifera sahariensis, A. m. intermissa, A. m. syriaca, A. m. lamarckii, A. m. ligustica, A. m. carnica*) populations (Shaibi *et al.*, 2009), as well as on the floral visitation patterns of bees in both agricultural (Bendifalla *et al.*, 2013) and natural habitats (Louadi *et al.*, 2007). The available information on diversity indicates highest species richness in the Mediterranean and Pre-Saharan areas (Figure 1), showing an unusual latitudinal gradient where diversity decreases southwards.

The bee diversity in the Eastern Mediterranean and Palearctic Middle East (Figure 1) is generally well documented although still having large gaps in knowledge (Grace, 2010). Diversity hotspots were identified for the Moroccan and Tunisian parts of the Atlas mountains, Libya, the Nile and Jordan valleys and Oman hills (Patiny & Michez, 2007). From the single taxon studies, none showed a preference for harsh desert environments and populations were instead concentrated in four main areas: 1) mountains extending within the Sahara and Arabic deserts – viz. Atlas Mountains, northwestern Libyan plateau and the western chains in Saudi Arabia and Yemen; 2) mountains dotting the central desert part of Sahara and Arabic deserts – Hoggar, Tibesti, and Tuwayq; 3) streams and river valleys – the Nile, the Jordan and others; 4) the coastal plains (Patiny & Michez, 2007; Patiny *et al.*, 2009).

Bee diversity in Eastern Africa (Figure 1.1) has been studied fairly well in certain countries. Kenyan bee faunas have been studied in natural habitats, those with different levels of disturbance and on cultivated and wild crops (Chiawo et al., 2017; Gikungu, 2006; Gikungu et al., 2011). The agricultural studies highlighted that even though A. *mellifera* was the main bee pollinator, crops also benefited from the visits of other wild bees such as *Xylocopa*, halictid bees and *Hypotrigona*, and that bee diversity increased with the increase of wild plant floral resources (Karanja *et al.*, 2010; Morimoto *et al.*, 2004; Njoroge *et al.*, 2004). In natural Kenyan landscapes, the dominant families were Apidae, Halictidae and Megachilidae, with Colletidae being scarce and sparsely distributed (Gikungu, 2006). Apifauna diversity was also found to vary seasonally in the eastern part of the continent. For rare species, long rains (March-May), dry season (December-February) and cold dry season (June-August) show similar levels of diversity while short dry season (September-October) had the lowest diversity (Gikungu *et al.*, 2011). For common species, the short rainy season (October-December) had the lowest bee diversity. Also interesting to note is that specific habitats were important for bee diversity in particular seasons, while other habitats were more or less stable all year round (Chiawo *et al.*, 2017; Gikungu *et al.*, 2011). A recent study in Mount Kilimanjaro, Tanzania, revealed bee richness patterns to be shaped by temperature and resources availability, presenting a continuous decline of bee species richness with increasing altitude, as well as a decrease in bee-flower interactions at higher elevations hence constraining resource exploitation (Classen *et al.*, 2015).

Overall, the bee fauna of East Africa is still poorly-sampled, probably due to few bee taxonomists and undersampling, although Kenyan museums have fragmentary bee holdings collected by naturalists or ecologists while sampling other targeted insect groups (Gikungu, 2006). Political unrest and infrastructural barriers in the Central region and the Horn of Africa have exacerbated the situation.

Bee research in Southern Africa (Figure 1.1) has intensified in the last two decades, fundamentally driven by and linked to pollination management interests. Most of the focus has been in South Africa, particularly in the Cape region 2004). The Honeybee races *A. m. scutellata and A. m. capensis* have been extensively studied in South Africa (Dietemann *et al.*, 2006; Hepburn & Guye, 1993; Johannsmeier *et al.* 1997). A species catalogue by Eardley & Urban (2010) provide a comprehensive bibliography of non-*Apis* studies. The bee diversity of Southern Africa is reasonably rich as it holds six of the seven bee families recognized by Michener (2007). Apifauna endemism is considered high in southern Africa (Figure 1), with the greatest numbers of endemic species present in the winter rainfall areas in the west and early mid-summer rainfall areas in the east of the country. Of the 89 genera occurring in southern Africa, 11% are endemic and of the 174 subgenera in the region, 20% are endemic, with the majority of endemics belonging to Megachilidae (30% of genera and 54% of subgenera) and Melittidae (40% of genera and 26% of subgenera) (Kuhlmann, 2009). Bee diversity in southern Africa is remarkably high exhibiting about 50% (1,400) of all known Afrotropical bee species (Kuhlmann, 2009). The Cape Floral Kingdom, situated in semi-arid to arid environment winter rainfall biomes, represents both a center of bee diversity and a hotspot of plant diversity (Gess & Gess, 2014; Kuhlmann, 2009). Michener (1979) suggested that the diverse topography and climatic contrasts of South Africa might explain the high levels of diversity of the regional bee fauna. The biotic distinctiveness of the region is entirely corroborated by the results obtained from national atlases of different taxa (plants, birds, frogs, reptiles and butterflies; Colville *et al.*, 2014).

The genera with highest number of species are *Lasioglossum* and *Megachile*, although this number is not particularly high when compared to other biodiverse regions. This apparent low species diversity in southern Africa may be an artifact resulting from the disparity between the size of the region and the number of African bee taxonomists, suggesting the possible existence of a large number of undescribed species or even new genera (Eardley *et al.*, 2009). The number of bee species is asymmetrically distributed among the families as Halictidae and Megachilidae are the most speciose, followed by Apidae and Colletidae, while Melittidae and Andrenidae are the least diverse (Kuhlmann, 2005; Kuhlmann, 2009).

A comprehensive study on South African bees revealed a clear association between climate (rainfall regime) and distribution patterns (Kuhlmann, 2009). The bee fauna was partitioned to species mainly restricted to either the winter rainfall parts (46,3% of species examined) or early to mid-summer rainfall (36,5% of species examined) with only a small number (18,9% of species examined) occupying the region with all year round rainfall (Gess & Gess, 2014; Kuhlmann, 2009). There were exceptions e.g. the large carpenter bees (*Xylocopa*) were more closely distributed in relation to temperature and altitude, with many species confined to the sub-tropical parts and decreasing diversity seen at higher latitudes (Eardley *et al.*, 2009).

The special climatic conditions, topography and geology of the landscape, and relative isolation of the island of Madagascar (Figure 1.1) makes this a unique case since approximately 90% of the 244 described bee species are endemic. The bee fauna in Madagascar is dominated by the families Halictidae (123 species) and Apidae (82 species) and nine of the 10 endemic genera belong to these families. The distributional patterns range from almost ubiquitous, to very restricted or even single locality collections (Eardley *et al.*, 2009; Kuhlmann, 2009; Pauly *et al.*, 2001).

Taxonomic work on bees in west and central Africa (Figure 1.1) is limited, although these regions are considered important centers of biological diversity (Rodger *et al.*, 2004). The absence of such data in these and many other African countries makes it difficult to undertake applied and pure pollination research in those areas, and there is a reliance of such work being carried out by foreign taxonomists (Brown & Paxton, 2009). The African bee fauna is also highly important given the likelihood of bees having originated from the continent (Danforth *et al.*, 2006).

The existing lack of basic knowledge on African pollination systems, particularly at community levels, limits interventions that address sustainable management of pollinators for both agriculture and conservation (Rodger *et al.*, 2004). It is estimated that once Africa is comprehensively surveyed 'the total number of bees in the Afrotropical region is likely to reach 3500-4000 species' (Kuhlmann, 2009).

1.3. BEE DIVERSITY IN ANGOLA

Angola (Figure 1.1) still constitutes a gap in the knowledge on insect biodiversity of Southern Africa (Kuedikuenda & Xavier, 2009). Typically, studies on bee diversity in the Afrotropical region or Southern Africa habitually omit Angola from the analyses due to the scarcity of information on the bee fauna of the region (Eardley *et al.*, 2009; Kuhlmann, 2009).

To understand the extent and tempo of bee research in Angola requires some consideration of both the history of the country and the prevailing socio-economic circumstances. Angola was a Portuguese overseas territory from the 16th century until its independence in 1975. Scientific expeditions to Angola were coordinated and promoted by the Board of Geographical Missions and Colonial/Overseas Research (Junta das Missões Geográficas e de Investigações Coloniais do Ultramar, or JIU) and constituted 'a way of assuring the success of the colonization process and the cost-effective exploitation of colonial resources, bringing obvious economic dividends for' Portugal (Castelo, 2012).

Specimens collected on the scientific expeditions were mainly sent to Portugal and can still be found in the leading collections, e.g. Institute for Tropical Sciences Research-IICT, in Lisbon, or the Museum at the Coimbra University, whilst some duplicates were left in Angolan collections. The collection in the Angolan Dundo Museum was entirely subsidized by Diamang - Companhia de Diamantes de Angola, a diamond exploration company founded in 1917 that installed a biological sciences laboratory within its headquarters, in the 1950's (Ribeiro, 1973). Most of the research on Angolan bee fauna was devoted to beekeeping and conducted between the 1950's and 1975.

After the national independence in 1975, Angola was affected by a civil war lasting for twenty-seven years (1975-2002) with the associated military interference that compromised infrastructure and scientific research. Additionally, financial constraints during and post war also contributed to the fact that almost no accurate data are available from 1975 onwards, for all biological fields including entomology and bee fauna in particular. The need to document the diversity and abundance of bee fauna in Angola is therefore urgent. Currently there is little published on the bee fauna of Angola, and certainly nothing on the composition and distribution patterns of the fauna.

Of the four known natural history collections in Angola, three contained entomological specimens and were affected, directly or indirectly, by the armed conflict. The entomological collection formerly housed at the National Museum of Natural History (MNHN) in Luanda was completely destroyed, not by the war itself but by the lack of maintenance during and after the period of conflict (scientific director of MNHN, *pers. comm.*, 2016). The collection housed at Dundo Regional Museum, Lunda Province, once considered one of the prime entomological collections within the Afrotropical region (Miller & Rogo, 2001), is still not accessible to the public or even to the scientific community and the preservation status of the collection remains unknown. The Agronomical Research Institute in Huambo, houses an entomological collection with *ca* 65,000 specimens, kept safe from military interference by the former Curatorial Assistant Francisco Elias (1951-present), now retired. Presently, the collection is housed at its original location, a colonial building without maintenance for the past 40 years, with no curator and a single untrained staff member responsible for preserving the specimens. Most of the Hymenoptera specimens in the collection, including bees, are not identified (Francisco Elias, *pers. comm.* 2016). This collection is currently being digitized under a BID-GBIF project entitled *'Strengthening the institutional network in Angola to mobilize biodiversity data'* (Elizalde, 2018).

Presently, the apifauna list for Angola includes 127 described species, from 37 genera and 4 families, from 928 existing records in the Global Biodiversity Information Facility (GBIF) platform distributed geographically according to Figure 1.3.

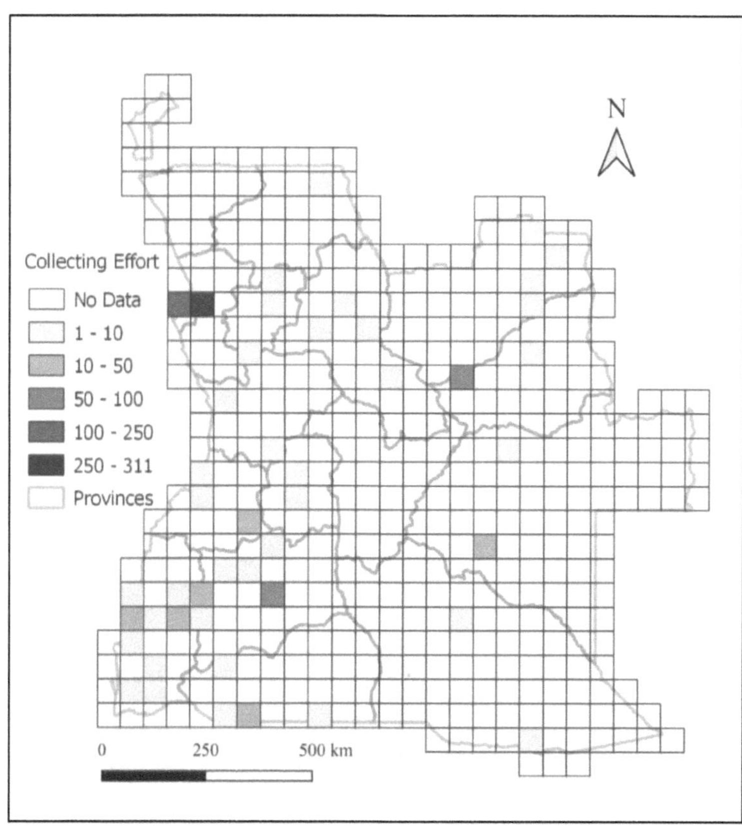

Figure 1.3 – Distribution of the collecting effort of Angolan bees' species as indicated by number of records per half degree grid cell, according to the dataset downloaded from GBIF. The dataset aggregates records from seven different natural history collections around the world. Cells in white lack records.

The vast majority (~ 78,3%) of the half degree cells holding records from bee species have a reduced (1-10) number of records. A smaller percentage of cells (13%) has up to fifty records and only a limited percentage (~ 9%) has more than fifty. The distributions of the available records reveal a strong collector bias, with higher sampling intensity around big cities such as Luanda, the capital. The collecting effort in the southwestern region of the country is better than that of the north-eastern area, but the distribution of records also clearly indicates a gap in knowledge on the strip in the east that extends to the southeast. The diversity of bee species in Angola will be further explored in chapter 2.

Altitudinal gradients have long been used by ecologists as 'natural experiments for testing ecological and evolutionary responses of biota to geophysical influences' (Körner, 2007). Factors such as climate zone, geographical region, area, net productivity and geometric constraints are known to influence spatial variation in species richness (Nogués-Bravo *et al.*, 2008).

Biodiversity distribution along altitudinal gradients has been mainly characterized by three patterns: i) monotonic decline of biodiversity with altitude (declining); ii) a low plateau where species richness is high until the mid-altitudes before declining (low plateau); and iii) a unimodal distribution with a mid-elevational hump (midpeak) (Grytnes & MacCain, 2007). The difference in patterns is related to the sensitivity of the studies to effects of area, sampling regime, sampling effort (Rahbek, 1995) and scale (Rahbek, 2005).

Insects, in particular, are strongly influenced along elevation gradients by environmental factors such as temperature, precipitation, wind turbulence, oxygen availability and radiation intensity (Hodkinson, 2005). The responses to changes can be direct, e.g. increased polymorphism in wing size, or indirectly mediated by interactions with other organisms, i.e. host plants, competitors, parasitoids or predators (Hodkinson, 2005). Depending on the biology of the species under consideration, the relationship between size and altitude can be positive or negative (Hodkinson, 2005). Because host plants also change along altitudinal gradients, insects' food consumption, food conversion efficiency, growth rates, survival and fecundity will all adjust correspondingly (Hodkinson, 2005). Population densities of insects show different responses to altitude as they can increase, decline or show no trend along the gradient, depending on the species (Hodkinson, 2005).

In temperate regions, studies have shown that altitudinal diversity patterns are season dependent. Moth species diversity decreased with increasing altitude in the Swiss Alps during spring and autumn, while in summer the diversity had a unimodal peak at mid-elevations and even though temperature may play an important role in this pattern, no explanation was found for it (Beck *et al.*, 2010). On the German Alps, Hoiss *et al.* (2012)

reported a linear decrease in bee species richness but increased phylogenetic community clustering with increasing elevation. The results from this study indicate that the proportion of social and ground-nesting species increase with increasing elevation in temperate regions, suggesting that community assembly at high altitude is influenced by environmental variables while competition processes may regulate the community structure at low altitudes (Hoiss *et al.,* 2012). Pollinator communities in temperate zones have been observed to display a significant change with altitude as anthophilous flies dominate the pollination provision at high altitudes and butterflies and bees dominating at low altitudes, which results in restricted outbreeding for some plant species (Hodkinson, 2005).

In tropical regions, where the influence of altitudinal gradients on diversity patterns have been well studied when compared to temperate zones, different patterns describing species richness along elevational gradients have been defined, including: a hump-shaped peak of species diversity at mid elevations caused by mid-domain effect, found in long term insect sampling studies (Brehm *et al.,* 2007); and b) increasing species richness with increasing altitude, found for galling insects in Brazilian high-altitude grasslands (Coelho *et al.,* 2017), parasitoid wasps in the neotropical region (Vejalainen *et al.,* 2014) or bee species diversity in Indonesia (Widhiono *et al.,* 2017). However, the most common pattern is a decline in plant and invertebrate species richness with increasing altitude, observed in eusocial wasps in Costa Rica (Kumar *et al.,* 2007); in bee and wasp communities in a Brazilian tropical mountainous region (Perillo *et al.,* 2017); or in natural and disturbed habitats in Mount Kilimanjaro (Classen *et al.,* 2015). The following factors have all been considered as driving forces influencing a linear decline of species richness along elevation gradients: increasing rainfall (Devoto *et al.,* 2005), competition between species (Hoiss *et al.,* 2012), climate and habitat management (Hoiss *et al.,* 2013), development time (Kocher *et al.,* 2014), habitat preference and behaviour (Koski & Ashman, 2015), reduction in habitat area and resource diversity (Fernandes et al., 2016), temperature (Peters *et al.,* 2016) as well as other physical phenomena as precipitation, water vapour pressure and wind speed (Perillo *et al.,* 2017).

Overall, temperature seems to be a main contributor for bee species richness and variation in life-history traits, both in temperate and tropical environments (Hoiss *et al.,* 2012; Classen *et al.,* 2015). Bees are expected to suffer strong selection pression at high altitudes due to the cooler conditions, experiencing physiological changes in life-history traits such as body size or coloration in order to improve thermoregulation (Hoiss *et al.,* 2012; Peters *et al.,* 2016).

Due to its general interest, life-history traits along elevation gradients have been widely investigated within single taxa, such as the grasshoper *Omocestus viridulus* (Berner *et al.,* 2004) in Switzerland or the allodapine bee genus *Exoneura* (Cronin, 2001) in Australia, but also at the community level, mainly investigating bees body size variation with altitude in temperate (Hoiss et al., 2012; Peters *et al.,* 2016) and tropical zones (Schellenberger-Costa *et al.,* 2017; Classen *et al.,* 2017).

Patterns in body size variation along elevational gradients, both inter- and intraspecific, are of great significance to evolutionary ecology and the most popular generalization is that body size of both endo- and ectothermic organisms increases with altitude (Hodkinson, 2005), a variation of the well-known Bergmann's rule that based on the concept that ectothermic organisms with larger body mass would lose smaller amounts of energy as their surface-to-body volume ratio was smaller (Brehm & Fiedler, 2004).

Bergmann's rule has been supported by research on birds and mammals, where the majority of the studied species increased in body size with increasing elevation (Meiri & Dayan, 2003), but failed to be demonstrated for freshwater fishes (Fu *et al.,* 2004) and ambiguous results have been obtained for amphibians, which displayed patterns of both increasing (Liao & Lu, 2012) and decreasing (Cvetkovic'*et al.,* 2008) body size with elevation.

Research on the variation of body size with altitude in insects, does not seem to follow a regular pattern among species belonging to the different insect orders. Coleoptera species have been shown to increase body size with increasing altitude (Smith *et al.,* 2000; Chown & Klok, 2003) but also to decrease (Chown & Klok, 2003). Hymenoptera species present increase in body size with increasing elevation (Ruttner *et al.,* 2000) but when investigated at the intraspecific level within communities showed a pattern of decreasing body size with increasing altitude (Classen *et al.,* 2017). Evidence that Orthoptera species support Bergmann's rule has been found (Rourke, 2000) as well as the opposite trend of declining body size with altitude (Bidau & Marti, 2007), to mention only a few. Moreover, other insect species show no significant variation in body size along altitudinal gradients, such as some species of geometrid moths (Brehm & Fiedler, 2004) or Coleoptera species (Chown & Klok, 2003).

Overall, body size is considered a complex life-history trait and its variation seems to be determined by a range of factors, including temperature (Kingsolver & Huey, 2008), growth rate and the mass gained during the developmental period (Davidowitz *et al.*, 2004). However, temperature (Angilletta et al., 2004) and seasonality of resource availability (Chown & Klok, 2003) seem to be the main drivers of this variation. Species with life-history traits of this level of flexibility are potentially sensitive to environmental disturbances, most probably leading to shifts in community composition or even species losses as a result of environmental change, affecting the functionality of the entire ecosystem (Petchey & Gaston, 2006).

Limited research has been applied to the effects of body size and other ecological traits of bee responses to anthropogenic disturbance, but the limited literature reveals that body size inconsistently affects the way species react to environmental changes (Williams *et al.*, 2010). In temperate zones, the response of bee body size to increasing altitude supports Bergman's rule (Maad *et al.*, 2013; Peters *et al.*, 2016; Scriven *et al.*, 2016). In tropical zones, the individuals body size variation has been reported to increase with increasing altitude while the overall body size within communities has been reported to decrease with increasing elevation (Classen *et al.*, 2017), therefore the difference between single species and community patterns suggests that there is a suite of smaller species at higher elevations.

Environmental variables associated with altitude, such as precipitation, have also been found to indirectly impact bee body size, particularly an augment with higher rainfall levels due to the increased food availability (Peruquetti, 2009).

1.5.   BROAD AIMS AND OBJECTIVES

The broad aims of this study are to i) consolidate the knowledge on Apoidea (Anthophila) diversity and distribution in Angola and 2) to examine the influence of a tropical altitudinal gradient (Bruco pass - Serra da Chela, Angola) on bee diversity and body size.

The specific aims and objectives are:

i.   To provide a profile of bee diversity in Angola, using existing databases, and use this to map distribution and richness patterns. This will identify gaps in collector effort and will broadly define zones of high species richness (Chapter 2).

ii.   To examine bee species composition along a tropical altitudinal and rainfall gradient and test whether patterns of decreasing species richness and abundance with altitude as recorded for temperate regions are true of tropical gradients (Chapter 3).

iii.   To investigate whether bee body-size increases with altitude along a tropical altitudinal and rainfall gradient, as has been reported for temperate altitudinal gradients (Chapter 4).

iv.   To compare the apoidean diversity patterns of Angola with those of other African regions (Chapter 5).

1.5.1.   RESEARCH HYPOTHESES RELATING TO THE ALTITUDINAL AND FAINFALL GRADIENT STUDY

1.   There will be no clear relationship between bee species richness and altitude as environmental contrasts will be less pronounced at different altitudes. This hypothesis is in contrast to trends shown in studies done in temperate (Hoiss *et al.*, 2012) and tropical (Perillo *et al.*, 2017) zones, where clear declines in species richness have been observed along altitudinal gradients.

2.   Bee body size will not show any variation with increasing altitude, particularly will not increase with increasing altitude. Similarly, this hypothesis contrasts with trends in temperate zones and tropical zones where positive relationships between bee

body size and altitude have been reported given altitudinal gradients in temperature, likely to be less pronounced in the gradient in this study (Peters *et al.,* 2016; Classen *et al.*, 2015).

3. There will be changes in bee community composition along the tropical altitudinal gradient altitude as rainfall varies along the gradient, influencing the vegetation communities and therefore (indirectly) affecting bee community assemblages, particularly for the more specialized bee genera and species (Perillo *et al.*, 2017; Classen *et al.,* 2015, 2017).

CHAPTER 2. BEE DIVERSITY IN ANGOLA

2.1. INTRODUCTION

The majority of angiosperms (87.5%) are pollinated by insects and other animals (Ollerton *et al.*, 2011). Pollination by insects of both crops and wild plants is a life-support mechanism, providing vital human nutrition globally (Eilers *et al.*, 2011) and sustaining biodiversity and ecosystem services (Ollerton *et al.*, 2011). Among insects, bees (Hymenoptera: Anthophila) are the most specialized and essential pollinator group across most ecosystems, thus playing an important role in ecosystem function (Waser & Ollerton, 2006).

Bees are a monophyletic group within the superfamily Apoidea with over 20,000 now described species worldwide belonging to seven recognised families: Apidae, Megachilidae, Andrenidae, Colletidae, Halictidae, Melittidae and Stenotritidae (Michener, 2007). This insect group is morphologically adapted to collect, manipulate, carry and store flowering plant products (Thorp, 2000), and some even specialize on particular host plants as a food resource, recognizing them by their unique flower scents (Milet-Pinheiro *et al.*, 2013).

Nearly all bee species rely on angiosperms products such as pollen and nectar to nourish adults and larvae (Nicolson, 2011; Scofield & Mattila, 2015). Nectar is the main source of carbohydrates, and pollen contains proteins, lipids and micronutrients essential for bee nutrition (Vaudo *et al.*, 2015). Other highly specialized bee species use energy-rich floral oils as a substitute (or in addition) to nectar for their larval or adult nutrition and water-resistant cell lining (Buchmann, 1987). Floral perfumes and waxes of angiosperms are used by bees as sexual attractants (Weber *et al.*, 2016) and resins are used to maintain the homeostasis of nest environment and reduce microbial growth (Simone-Finstrom & Spivak, 2010, 2012).

Generally, the knowledge on African pollination biology is still very limited (Rodger *et al.*, 2004), as the continent is poorly surveyed and under-studied with most countries facing numerous barriers to research, including but not limited to lack of taxonomists, lack of infrastructure and political instability (Eardley *et al.*, 2009).

Most of the research conducted in Southern Africa has been extensively compiled into a species catalogue (Eardley *et al.,* 2009; Eardley & Urban, 2010) and is predominantly dedicated to taxonomic revisions and a few descriptions of new species. A compilation of information on bee biology and how land use impacts their diversity was also produced for this region of the continent (Gess & Gess, 2014). Limited work on ecology and social biology of bees has been conducted in South Africa, both on fruit production (Carvalheiro *et al.* 2010; 2012) and in crop production (Allsopp *et al.,* 2008; Carvalheiro *et al.,* 2011; Melin *et al.,* 2014), with the focus on the Cape region. The studies in this region mainly conclude that fruit and crop productivity are enhanced by the co-existence of patches of natural habitat and floral diversity (Carvalheiro *et al.,* 2011), contributing towards food security, as well as biodiversity conservation on the continent. The patterns of diversity, endemism and distribution have also been analysed, showing a pattern of highest species diversity occurring in the arid west, and in the relatively moist east extending to the winter rainfall region (Eardley, 1989; Kuhlmann, 2005; 2009).

In the Malagasy region, Madagascar is considered a global conservation priority due to its extraordinary biodiversity and high level of endemism. The country faces high levels of deforestation and habitat fragmentation and the preservation of the endemic and ecologically important species is highly dependent on forest preservation (Goodman *et al.,* 2003). A comprehensive work on Madagascar's Anthophila fauna, generated a checklist of 244 bee species, with the description of 3 new genera and 62 new species (Pauly *et al.,* 2001). This work also concluded that the conservation of natural habitat, especially the humid forests and western dry forests, is essential to preserve the diversity of bees in this area of the continent (Pauly *et al.,* 2001).

In eastern Africa, studies on pollination, bee diversity and ecological aspects of their interaction with wild plants (Gikungu, 2006; Gikungu *et al.,* 2011) and vegetables (Morimoto *et al.,* 2004) have determined that applying strategic and holistic management to both natural forests and agro-ecosystems will contribute to biodiversity conservation and safeguard pollination services.

More recently, studies performed in the sub-Saharan drylands of west Africa reveal that insect pollination, especially bees, improve yield quantity and quality of economically important crops. These studies also conclude that habitat conservation is vital to protect bees, consequently assuring pollination services of fruit-trees and crops that are the basis of local livelihoods (Stein *et al.*, 2017; Stout *et al.*, 2018).

The Republic of Angola, located in the south-western coast of the African continent, is the largest country in southern Africa with a total area of 1,246,700 km², varying in elevation from 0 to 2620 m above sea level (Figure 2.1).

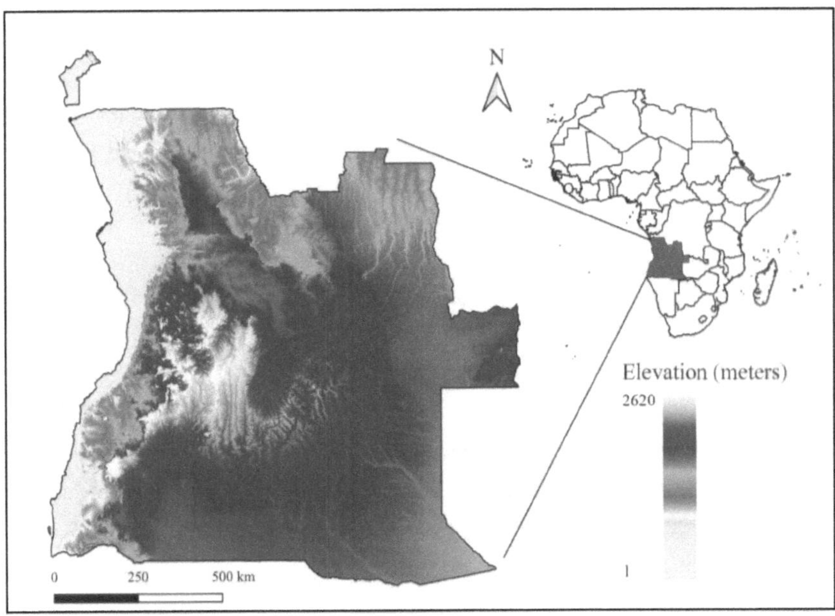

Figure 2.1 - Representation of the digital elevation model for Angola. Data source:

This southern African country includes a variety of climatic features corresponding to five climate types of the Koppen–Geiger system (Peel *et al.*, 2007; Figure 2.2), ranging from tropical humid in the north to extremely arid in the south west.

In terms of biogeographic units, 15 World Wildlife Fund (WWF) terrestrial ecoregions (Olson *et al.*, 2001; Figure 2.3) are recognised in Angola, of which the most widespread is the miombo woodlands occupying the Central Plateau and being replaced in the south by the next widespread unit, mopane woodlands (*Colophospermum mopane* and *Baikiaea plurijuga*). The north encompasses the Congolian forest savanna mosaics, scarp savannas and woodlands. The south-western coastal plain is characterised by Kaokoveld desert and Namibian dry savannah woodlands (Romeiras *et al.*, 2014). The Angolan phytogeography chart by Grandvaux-Barbosa identified 32 vegetation units, from rainforests to desert (Grandvaux-Barbosa, 1970).

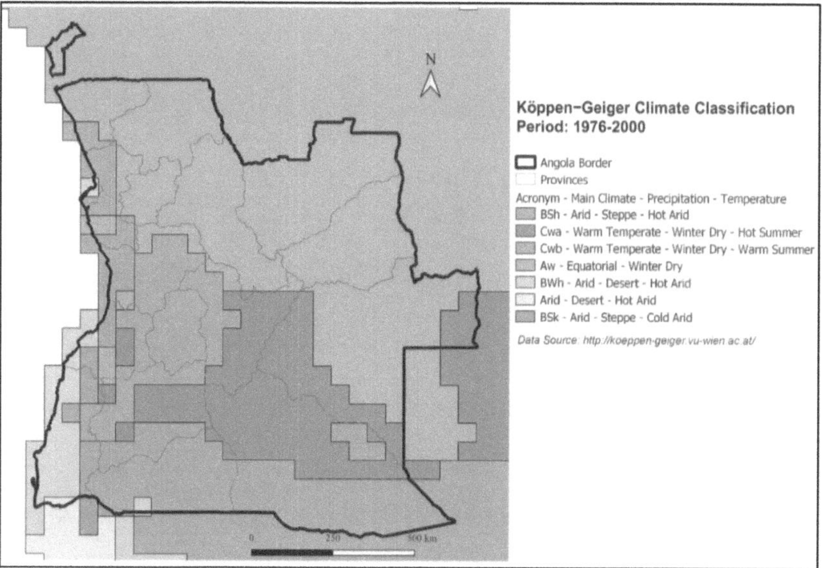

Figure 2.2 – Representation of the Koppen-Geiger climate classification system overlaying on Angola's borders.

Angola falls almost entirely within the Zambezian region (Linder *et al.*, 2012), with two exceptions: 1) In the north, where a transition scarp falls in the Congolian biogeographical unit (Shaba); and 2) in the south west and south, where transition edges fall into the Southern African unit (south western Angola and Kalahari, respectively) (Linder *et al.*, 2012).

Given that the Zambezian region is known for its high levels of species turnover (Linder *et al.*, 2012) and Angola has an incredible diversity of habitats within its boundaries, the country is expected to be 'one of the richest in species in Africa' (Kipping *et al.*, 2017). Despite the diversity of ecosystems and species richness in Angola, research in almost all biological fields is scarce (Figueiredo *et al.*, 2009; Kipping *et al.*, 2017; Mills, 2010; Rodrigues *et al.*, 2015; Romeiras *et al.*, 2014).

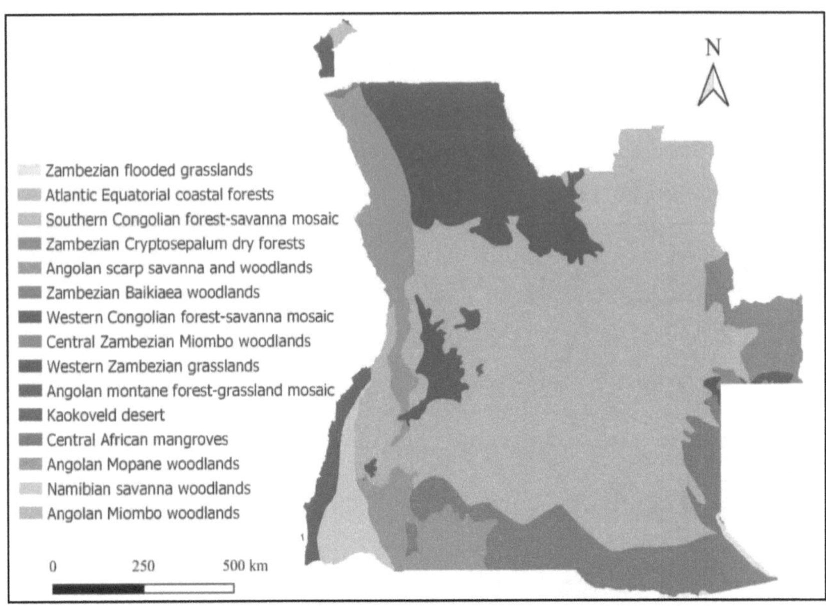

Figure 2.3 - Representation of the 15 recognized terrestrial ecoregions of Angola (adapted from Olson et al., 2001).

Until Angola's national independence in 1975, scientific research in the country was mainly conducted by overseas researchers and resulted in the collection of thousands of biological specimens, some kept locally but most sent to Portugal and other European countries, e.g. France, United Kingdom, Switzerland and Germany (Crawford-Cabral & Mesquitela, 1989; Figueiredo *et al.*, 2009). Existing collections such as those at Lisbon University, which inherited the specimens from Instituto de Investigação Científica Tropical (now defunct), the Natural History Museum in London, the Royal Botanic Gardens in Kew or the Muséum National d'Histoire Naturelle in Paris, to name only a few, are still rich in type material from Angola (Figueiredo *et al.*, 2009).

Broad-spectrum, natural history collections are reservoirs of baseline data on diversity, taxonomy and historical distributions of species. Integrated with spatial records, historical data can inform aspects of ecological and evolutionary theory, be applied to conservation, determine early records of invasive species or even harbour information on species that have become locally extinct (Graham *et al.*, 2004; Hoeksema *et al.*, 2011). Even though biological collections from that historical period were mostly arbitrary and opportunistic, they constitute a suitable source of information on species distribution or information on variation of individuals attributes to environmental variables (Pyke & Ehrlich, 2010).

The Angolan National Independence War (1961-1975), a revolution against Portuguese colonisers, that was immediately followed by the Angolan Civil War (1975-2002), a power struggle between the three main liberation movements (Guimarães, 1992), resulted in the exodus of researchers and destruction of facilities, obstructing field work and the study of local collections (Figueiredo *et al.*, 2009). The lack of researchers including taxonomists, as well as access to historical collections and infrastructure contributed for a generalised gap of biological research in Angola for over a period of more than forty years.

The diversity of the bee fauna in Angola is reviewed in this chapter with the main goals being: 1) to review and compile in a digital format the data on bees' (Anthophila) species distribution for Angola, as extracted from the literature 2) to review and compile in a digital format the information on bee diversity present in historical collections 3) to use the full data set to identify spatial gaps in knowledge.

2.2. METHODOLOGY

2.2.1 Geographical framework

The present work deals with the area of the Republic of Angola, located in the south west coast of the African continent.

2.2.2 Literature review on bees (Anthophila) diversity and distribution in Angola

An extensive survey on potential sources of biogeographical data for bees in Angola was undertaken, using online libraries of peer-reviewed journals as a starting point. Additionally, all physically available publications from *Junta de Investigações do Ultramar,* an organization created in 1963 to support research in the Portuguese colonies, were consulted. The process involved filtering journals with relevant information for this study present at Angolan libraries and collecting records from each publication. Some of these publications are also available on the data portal *Memórias de África e do Oriente*, a project of *Fundação Portugal-África* developed and maintained by *University of Aveiro – Portugal* and *Centro de Estudos sobre África e do Desenvolvimento*, since 1997

Another valuable source of information for the bibliographic survey were the publications from *Diamang – Companhia de Diamantes de Angola*, a diamond corporation occupying a total area of 52,000 km^2, that promoted the scientific knowledge on all biological fields, situated in the northeast of the country, currently Provinces of Lunda north and south (Ribeiro, 1973).

The publications from both *Junta de Investigações do Ultramar* and *Diamang* are the main sources of information on the research performed in Angola before the national independence in 1975.

Supplementary sources of information for bee species records for the country, or with accurate point locality data, were online projects such as WaspWeb Discover Life Atlas Hymenoptera the collection from the Smithsonian National Museum of Natural History and Global Biodiversity Information Facility under the following search parameters: Country: Angola| Basis of record: preserved specimen + unknown| Scientific name: Animalia + Arthropoda + Insecta + Hymenoptera + Apidae + Megachilidae + Halictidae + Andrenidae + Anthophoridae + Colletidae.

In order to avoid duplication of data (hence statistical independence of data points), all data available at the different data providers (collections) websites were cross-checked with the data downloaded through GBIF. The database SouthABees – 'Southern African Bee Data Base' (Huber *et al.*, 2005) was still not available during the timeframe of this work.

Scope of data retrieved from literature:

- *Area*: only records from locations within Angola's borders were included;
- *Time period*: the bibliographic review included all available literature from the colonial period until the present.
- *Data validation procedures*: The taxonomic names mainly follow the nomenclature in Eardley *et al.* (2010). All original names in sources were checked and cleaned when necessary. Species were classified as invalid if their identity could not be established, or if their validity was questionable (*nomen dubium*);
- *Location validity*: Individual records in the literature were checked, as to whether the original location was within Angola's borders, using old locality lists and maps. The gazetteers from Crawford-Cabral & Mesquitela (1989) and Mendes *et al.* (2013) were especially useful for validating the geographical information;
- *Record validity*: There was no exact way to check for the validity of the records. Being a literature review, it is assumed to have a wide range of different quality data and there was no reason to doubt the precision and validity of the records.

2.2.3 Natural History Collections: a revision of the information on bee diversity

From the three Angolan natural history collections containing entomological specimens referenced in Chapter 1, unfortunately only two were able to keep the specimens safe from military interference or armed conflict damage.

The efforts of Mr. Franscisco Elias, curatorial assistant at the Agronomic Research Institute (IIA) in Huambo, and Dr. Esteves da Costa Afonso, curator of the biological collections at the Dundo Regional Museum in Lunda Province, to maintain the collections for all these years and against all odds (Franscisco Elias and Esteves Afonso, *pers. comm.* 2017) are invaluable. During the timeframe of this work, I was only allowed to work with the collection in IIA – Huambo.

Existing bees collections held in Portugal with specimens from Angola, presently belonging to Lisbon and Coimbra Universities, have yet not been studied or digitised (Luis Mendes and António Gouveia, respectively, *pers. comm.* 2017), and for that reason there is no information available from these sources.

The two most representative entomological collections in Europe, from the Natural History Museum of London (NHML) and the Muséum National d'Histoire Naturelle in Paris (MNHNP), are still not entirely digitized. The entomology collections of NHML hold 34 million insects and arachnids of which only about 4% (1,333,982) have been digitized and are available online. None of the digitized records belonged to bees (Natural History Museum, 2018). The Hymenoptera collection in the Muséum National d'Histoire Naturelle in Paris, that holds about 1 million specimens of which 210,000 (21%) are bees (Anthophila), is still not fully digitized but in the Afrotropical zone the main geographic area represented is Madagascar (Guiraud & Chagnoux, 2018).

2.2.4   Statistical and Spatial Analysis: Identification of the gaps in knowledge

All locality data with point source records found in the bibliographic revision were mapped to assess patterns of collecting effort and generic richness of bees (Anthophila) in Angola.

The completeness metrics were computed using EstimateS (Colwell, 2013; version MAC 9.1.0) and included: the number of records per grid cell, the Abundance-based Coverage Estimator (ACE), the CHAO1 completeness index (Gotelli & Chao, 2013), and generic accumulation curve for each grid cell. The generic accumulation curve derived by EstimateS is based on an average of a series of 100 randomizations of the genera by grid data matrix, sampling with replacement, at each level of abundance and typically starts rising rapidly, representing the observation of common species, and continues rising slowly requiring more sampling to detect all rare species (Colwell, 2013; Gotelli & Chao, 2013). This randomization allows for the independence of the generic accumulation curve from the order in which the records were added to the analysis.

The availability of biodiversity data mainly through open-access databases is essential for research and conservation, but its effectiveness is highly dependent on survey completeness (Troia & McManamay, 2017). Following Coddington *et al.* (2009), a grid cell was considered as 'well-surveyed 'when containing ≥ 10 occurrence records and if completeness was higher than 75%; completeness was calculated by dividing the number of records by the estimator ACE, per grid cell, following the equation:

$$Survey\ completeness = \left(\frac{Sobserved}{Sestimated}\right) x100$$

To identify the gaps in knowledge, all occurrence records and survey completeness were mapped using QGIS version 3.4.2, and a grid of half-degree cells was superimposed onto the polygon of the country with the limits of provinces clearly marked.

2.3.   RESULTS & DISCUSSION

2.3.1   Preliminary checklist of bee species for Angola

The zoological expeditions to Angola that enriched natural history collections all over the world, date back to the 18th century, starting in 1784 and proceeding for almost 200 years, until 1974 (Crawford-Cabral & Mesquitela, 1989). The available literature from these expeditions reveal that researchers were focused in collecting specimens from the following fields of research: botany, mammalogy, ornithology and herpetology. For other insect groups (e.g. Odonata), research began in 1928 on the first expedition led by Albert Monard, a swiss zoologist curator at the Natural Museum of La-Chaux-de-Fonds. Following the tendency, Monard mainly collected vertebrates and plants on his two expeditions to Angola, but because of his broad interest in nature he also collected (at least) some Odonata specimens (Kipping *et al.,* 2017). Only the Tring Museum Expedition in 1934 was led by an entomologist, Karl Jordan, and even though mainly vertebrates were collected (Crawford-Cabral & Mesquitela, 1989), given his research interest, it would be expected that he collected Angolan entomological specimens and that these would be held in this British museum. The Tring Museum is part of the Natural History Museum of London and its collection from Angola still remains to be digitized.

The bulk of Angolan research on bees during the colonial period was dedicated to beekeeping and started in the 1950's. Most of these publications are not available in Angolan libraries, but should be available at the National Documentation Centre, located in the capital, Luanda. Also, copies of these are expected to be found, at least, on libraries in Portugal such as the one from the former IICT. A complete list of the publications that could not be physically located in the visited libraries in Angola is presented on Appendix I.

In this period of the history, Angolan beekeeping seemed to be based on three sub-species of *Apis mellifera*: *Apis mellifera adansonii* Latreille, *Apis mellifera unicolor* Latreille and *Apis mellifera intermissa* Butt-Reep. Stingless bees were also found as a potential source of honey with many advantages, especially their harmless nature. In the 1950's three species were used in Angola for cultures: *Meliponula bocandei* Spinola, *Meliponula togoensis* Stadelman and *Trigona (Hypotrigona) gribodoi* Magretti. Additionally, *Cleptotrigona cubiceps* Friese was being studied in culture because even though it had no value as honey producer, it was thought to be a robber bee of *Trigona (Hypotrigona) gribodoi* (de Portugal Araújo, 1957).

Other research topics studied in that period were related to sources of honey and native methods of beekeeping (de Portugal Araújo, 1955), methods of processing honey and its products, as well as characteristics of the African bee and its enemies (e.g. the honey-guide *Indicator indicator*) (Rosário-Nunes & Tordo, 1960).

Generally, the traditional bee-keeping methods used by local people were described as somewhat primitive as they collected the hives from holes in trees through the use of fire that destroyed the bee colony and the hives. The 'quiocos', a Portuguese word for côkwe (the côkwe ethnicity occupies all northeast Angola and a large strip on the east that extends to the south), were the only exception as they maintained their hives to ensure the continuity of the swarm exploitation (Rosário-Nunes & Tordo, 1960). References also mention that even though the honey was sometimes offered for sale as a food product, the methods used in its preparation and the resulting low level of purity, made it unattractive to European consumers. Local people would mostly consume the honey in fermented drinks (known locality as 'Hidromel', or in some areas as 'bingundo') and very rarely as food.

From 1957 onwards, several campaigns were undertaken by Portuguese beekeeping specialists, aiming to provide technical guidance to the indigenous people in order to improve their production skills and the quality of the final product. Numerous apiary outposts were constructed in the country using mobile hives. This project proved to be a challenge as, despite all the training, local people chose to keep using the traditional methods that involved tree barking to produce wood hives, which destroyed approximately a million adult trees per year, and the use of fire to collect the honey (Rosário-Nunes & Tordo, 1960). This inefficient and environmentally harmful methodology is still used today (Figure 2.4).

Figure 2.4 - Traditional wood hive found in Luando Natural Integral Reserve, Angola, August 2018.

Contrasting with the poor quality of the honey, the wax produced in Angola was highly sought after in international markets, and, therefore of economic importance. Minister Sá da Bandeira from the Huíla Province, in 1858, ordered improvements in the preparation of this product which led to the creation of the decree Portaria nº 2267, of 1937, containing guidelines of best practices to produce exceedingly valuable wax (Rosário-Nunes & Tordo, 1960).

In summary, although the honey itself as a product was not very valuable, its by-products were, and so the colonial government had an interest in promoting research mainly related to beekeeping.

Prior to the present revision, only 127 bee species were known from Angola (Ascher & Pickering, 2018) and the present work added 82 species to the country's checklist, increasing it by 64.6%.

All available literature on Angolan bee diversity, as well as online data organizers, were reviewed resulting in a preliminary checklist of 186 bee species recorded for the country, from 41 genera and 5 families as summarized in Table 2.1. This study allowed the addition of 23 species and 6 genera to this list (see chapter 3), elevating the national total to 209 species, 47 genera belonging to 5 families. The preliminary checklist of bees of Angola with information on sources is presented in Appendix II.

Table 2.1 – Summary of the number of bee species records from Angola found in literature and online data sources, aggregated per genera and family

Family	Genera	Number of species per genus	Family	Genera	Number of species per genus
Andrenidae	*Melliturga*	2	**Halictidae**	*Cellariella*	2
	Meliturgula	1		*Ceylalictus*	2
				Halictus	1
Apidae	*Allodape*	2		*Lasioglossum*	2
	Amegilla	16		*Lipotriches*	8
	Anthophora	7		*Nomia*	3
	Apis	1		*Pseudapis*	5
	Braunsapis	13		*Seladonia*	4
	Ceratina	7		*Sphecodes*	-
	Cleptotrigona	1		*Thrinchostoma*	3
	Dactylurina	1			
	Hypotrigona	3	**Megachilidae**	*Afranthidium*	1
	Liotrigona	2		*Anthidiellum*	1
	Macrogalea	1		*Anthidium*	1
	Meliponula	3		*Coelioxys*	3
	Pasites	1		*Euaspis*	1
	Tetralonia	3		*Heriades*	2
	Tetraloniella	1		*Megachile*	43
	Thyreus	7		*Noteriades*	1
	Xylocopa	23		*Pachyanthidium*	2
				Pseudoanthidium	1
Colletidae	*Colletes*	-		*Sterapista*	1
	Hylaeus	2		*Trachusa*	1

For the compilation of the preliminary checklist presented on Appendix II, the peer-reviewed Catalogue of Afrotropical Bees (Eardley & Urban, 2010) was used as dependable baseline and then supplemented with records from biological databases. Data aggregated on non-quality controlled databases (such as GBIF) is of variable completeness and precision, therefore needs filtering and post processing (Sikes *et al.*, 2016). The data were vetted against current taxonomy and localities verified.

Although insects represent a high percentage (ca. 75%) of species and specimens from museum collections, there has been relatively little effort made in digitizing this vast amount of data. In fact, in 2016 only 7% of the occurrence data available through the data organizer GBIF was entomological in nature (Sikes *et al.*, 2016).

The different data providers (natural history collections) consulted for this revision share their data through GBIF and for that reason this platform was the main source of information with point source records. After applying the relevant filters mentioned in the methodology section, 928 unique records were obtained. These records were aggregated from seven different data providers, the Snow Entomological Museum Collection from the University of Kansas being the biggest contributor (SEMC, 773 records).

Following the best practices of citing data obtained from data aggregators (Sikes *et al.*, 2016) and in order to acknowledge the data providers for their contribution to science, the links (DOI) for the data downloaded from GBIF are indicated in Table 2.2.

Table 2.2 – Links DOI for the data downloaded through GBIF from different natural history collections.

The importance of access to collection-based archives of biological data cannot be overemphasized (Krishtalka & Humphrey, 2000) and Angola is presently making efforts in using information technology to provide electronic access to its vouchered biotic information through a project named "Strengthening the institutional network in Angola to mobilize biodiversity data", funded by the European Union through a Biodiversity Information for Development (BID) grant from GBIF (Elizalde, 2018). The author of this book is one of the coordinators of this project.

Starting in July 2016, the project has collaborated with national institutions promoting the mobilization of biodiversity data through awareness actions, digitization of natural history collections, management and curation of recently obtained biodiversity data and several specific training workshops. As of November 2018, almost 11,000 records were digitized, georeferenced and published in the GBIF portal and about 40,000 are at the final stages of the quality control process to be published until the end of February 2019. During the project, more than twenty personnel from the different institutions benefited from several training sessions in digitizing collections, data sharing and georeferencing.

One of the project's main partner is the Agronomic Research Institute (IIA) in Huambo, a public institution since 1961, under the Ministry of Agriculture, devoted to research, technological development and innovation (Deputy-director, *pers. comm.* 2017). The IIA holds at its headquarters an herbarium and entomological collections, of national and international interest, mainly dating back to the colonial period and the first post-independence years. The herbarium with more than 40,000 specimens, is recognised as having high importance because it hosts duplicates of the Gossweiler collections, and also hosts duplicates of the holotypes (now isotypes) from Antunes and Dekindt, that were destroyed in the Berlin Herbarium during World War II (Figueiredo *et al.,* 2009). The entomology collections were initiated by Passos de Carvalho with agricultural purposes (representing pests and pollinators) but rapidly turned into a general entomology collection (Figure 2.5) with an estimation of more than 60,000 specimens, with valuable collections of Lepidoptera (about 15,000 specimens) and Orthoptera (about 10,000 specimens).

The best represented orders in the collection are Coleoptera (about 20,000 specimens), With the exception of Lepidoptera and Odonata, that were determined by the prominent entomologist Elliot Charles Gordon Pinhey, most specimens in the collection are either not identified or identified only up to family or genera level.

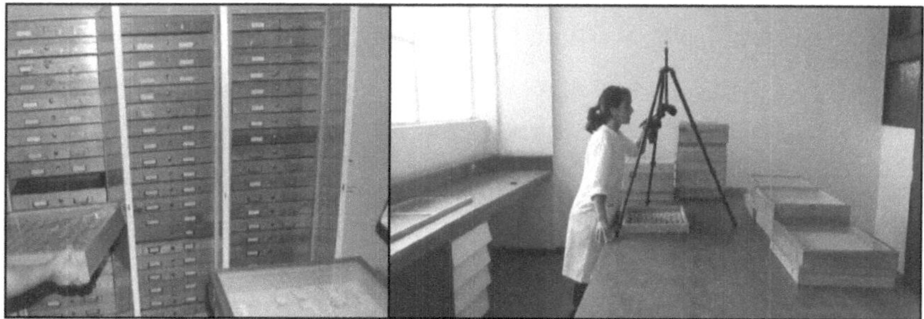

Figure 2.3 - Left: a detail of the collection room in the Agronomic Research Institute (IIA) in Huambo; Right: The institute has limited resources and the student improvised to take pictures of the pinned insects.

The Hymenoptera collection is composed of an estimated 6,000 specimens including wasps, ants and bees. The bees (Anthophila) collection held at IIA-Huambo has 1,436 specimens of which 396 are identified to species level, being mostly duplicates of six species: *Xylocopa caffra; X. flavorufa* (Figure 2.6); *X. combusta; X. inconstans; X. olivacea and Apis mellifera.*

Figure 2.6 - Image of the box nº 60 from the Agronomic Research Institute entomological collection with specimens of Xylocopa flavorufa De Geer

During the time frame of this work, I identified the remaining 1,040 specimens of the bees (Anthophila) collection to generic and subgeneric level following Michener (2007). All 1,436 records were digitized, georeferenced and the pinned insects were photographed in the boxes. By March 2019, these records will be publicly available online at the GBIF data portal, increasing the data on bees (Anthophila) from Angola currently available at this global platform by 154,7%.

2.3.3 Distribution patterns of Angolan bees

The large size of Angola, its diversity in habitat types and the halt in research caused by the armed conflict that lasted for almost three decades, are all factors contributing to the poor knowledge on the country's biodiversity, which is one of the least known in the African continent (Ceríaco *et al.*, 2014). Additionally, there seems to be a pattern of lack of collecting effort in the east of the country which is also true for other fields of biological research (Clausnitzer *et al.*, 2012; Taylor *et al.*, 2018).

Mapping the 928 unique records downloaded from the GBIF data portal resulted in a distribution of collecting effort that not unexpectedly revealed a collector bias around big cities, such as Luanda that holds 311 records of bee specimens (Figure 2.7). The collecting effort in the southwestern region of the country is better than that of the north-eastern area, but the distribution of records also clearly indicates a gap in knowledge on the strip in the east that extends to the southeast. The vast majority (~ 78,3%) of the half degree cells holding records from bee species have a reduced (1-10) number of records. A smaller percentage of cells (13%) has up to fifty records and only a limited percentage (~ 9%) has more than fifty.

The representation of the 1,436 specimens of bee species digitized from the Agronomic Research Institute collection in the Angolan polygon, also revealed a collector bias in the distribution of the collecting effort. The institute is composed of several experimental stations distributed through the country and the cells holding the higher number of records are coincident with the location of those stations. The institute headquarters in Huambo had the highest collecting effort with 522 records (Figure 2.8). Some records are also coincident with the main roads in the country, possibly indicating that collections were made when collectors moved between the experimental stations.

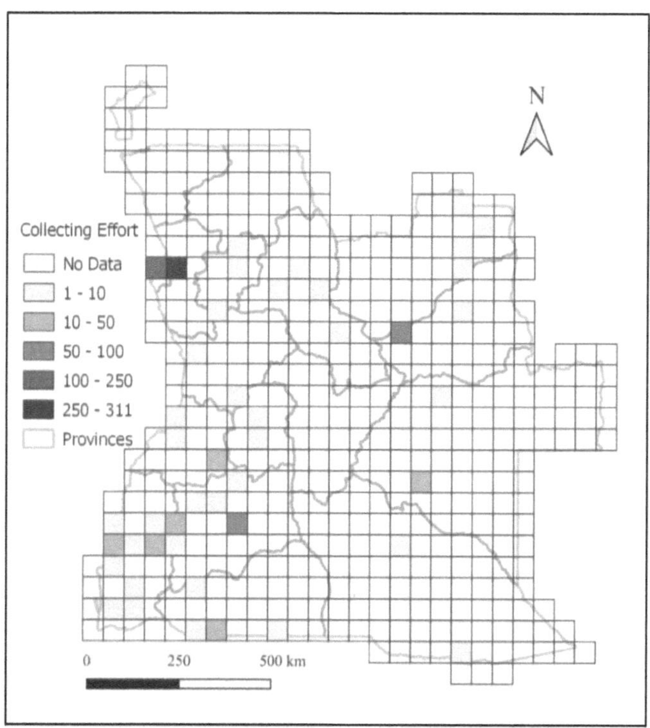

Figure 2.7 - Distribution of the collecting effort of Angolan bees species as indicated by number of records per half degree grid cell, according to the dataset downloaded from GBIF. The dataset aggregates records from seven different natural history collections around the world. Cells in white lack records.

Following the pattern, the entire eastern region and the north of the country are underrepresented in this natural history collection. The collection effort represents the number of specimens from each half degree grid cell as the collection is mostly identified to genera level.

The entomological collection from IIA will be uploaded to the GBIF platform soon and, for that reason, a combination of all records above mentioned will be available on bees (Anthophila) from Angola in the near future.

The combination of data clearly shows collector bias around big cities and specific locations, such as the agronomic experimental stations. The eastern half of the country is largely undersampled and the north western tip of the country seems never to have been surveyed (Figure 2.9).

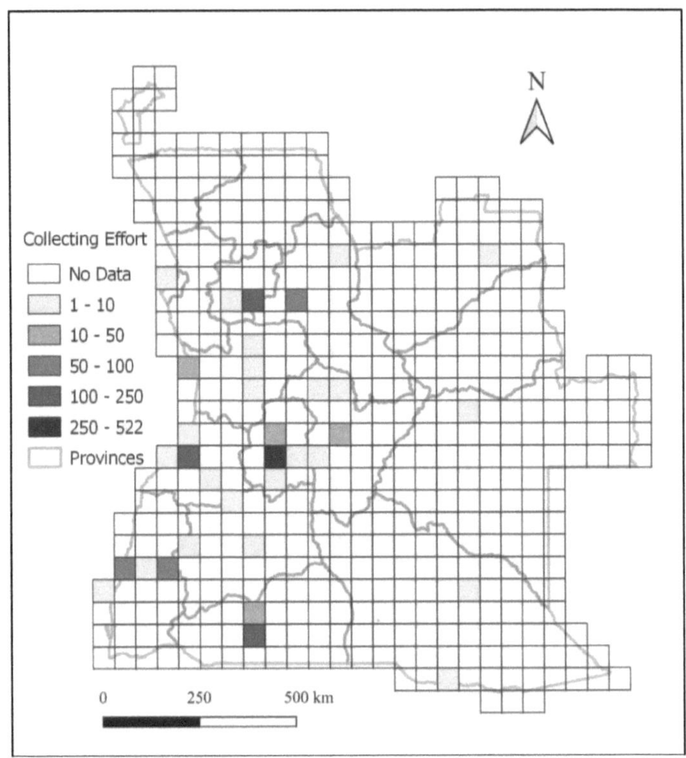

Figure 2.8 – Distribution of the collecting effort of Angolan bees as indicated by number of records per half degree grid cell, according to the digitization of the entomological collection from the Agronomic Research Institute in Huambo (IIA). Cells in white lack records.

It is important to note that of the 2,364 records, the vast majority (1,303; 55%) were collected between 1970 and 1975 by the researchers working for the Agronomic Research Institute. Before (and after) this period, only one other large collection (516 records; 21,8%) was registered in 1957. This collection is entirely housed at the Snow Entomological Museum Collection and its main collector was Virgilio de Portugal Brito Araújo, considered the Portuguese master of bees, who worked as a beekeeping technician for the Angolan Agriculture and Forestry until 1961 (Kerr, 1984). Given the similarity of collection sites, it is possible that Portugal Araújo partially joined the 1957 Gerd Heinrich Expedition, the second expedition lead by this German explorer in Angola from where zoological specimens were sent to the Peabody Museum (Crawford-Cabral & Mesquitela, 1989).

Since the Huambo's bee collection, that represents 60,7% of the data considered in this revision, was mainly identified to genera level, the geographical pattern of distribution was analyzed at this taxonomic level. As expected, the geographic pattern of genera richness followed the same pattern of the collection effort, being highest around big cities or the experimental stations location, with the highest number of different genera (23) recorded from the Huambo cell (Figure 2.10).

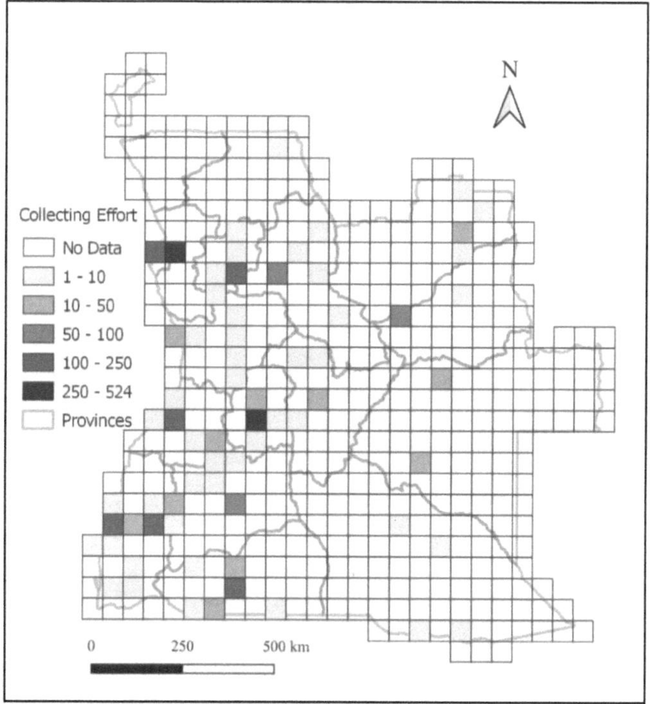

Figure 2.9 – Distribution of the collecting effort of Angolan bees as indicated by number of records per half degree grid cell, according to the GBIF dataset and the entomological collection from the Agronomic Research Institute in Huambo (IIA). Cells in white lack records.

The vast majority (~ 77,6%) of the half degree cells holding records have a reduced (< 5) number of different bee genera. A smaller percentage of cells (~ 13%) has up to ten different genera and only a limited percentage (~ 8%) has up to seventeen. Only a single cell presents a higher generic richness (23).

However, merely counting the number of genera in a sample will return a biased underestimate of the true number of genera as the number of observed genera will inevitably increase with increasing collection effort (Gotelli & Chao, 2013). This effect is best illustrated in a *generic accumulation curve*, where the number of collected individuals is plotted against the number of observed genera.

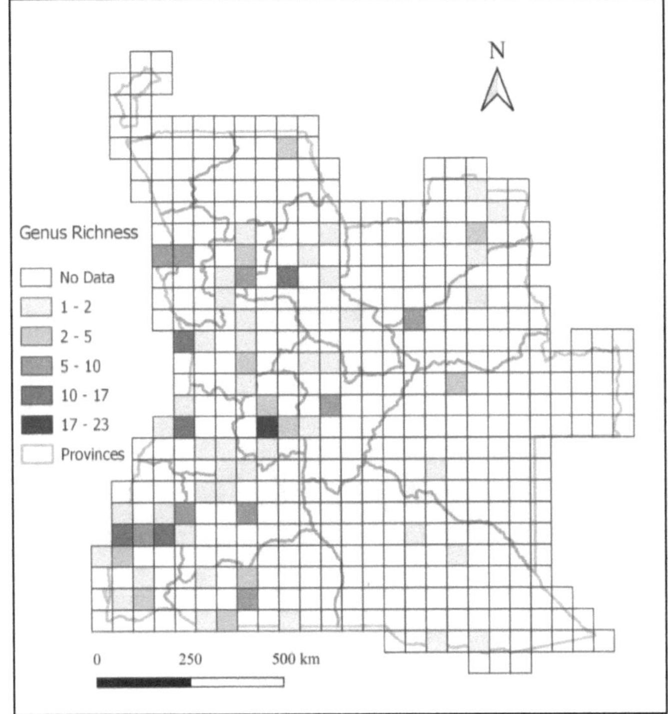

Figure 2.10 – Genera richness of Angolan bees as indicated by number of unique genera per half degree grid cell. Cells in white lack records.

Some of the results of the metrics computed using EstimateS for the different half degree grid cells presenting survey completeness ≥ 75% are summarized in Table 2.3 (grid cells named according to its location in the country).

The result for the Huambo grid cell was left in the table, even though survey completeness was only 55%, in order to exemplify the biased underestimation resulting from merely counting the number of records. The Huambo grid cell presents the highest number of collected specimens (523) and genera (22) but its survey completeness was calculated in 55%, while for the Luanda 2 grid cell holding 311 collected specimens and a richness of only 6 genera the survey completeness was calculated as 100%. As can be observed by the comparison of the generic accumulation curve for these two grid cells (Figure 2.11), the Huambo grid cell is highly undersampled.

Table 2.3 - Summary of the results from the computation of completeness metrics using EstimateS (Colwell, 2013) for some of the half-degree grid cells in Angola holding records of collected bee specimens

GRID-CELL NAME	NUMBER OF SPECIMENS COLLECTED	S MEAN	ACE MEAN	CHAO 1 MEAN	COMPLETENESS ACE	COMPLETENESS CHAO 1
NAMIBE	114	12	13,98	15,96	86%	75%
CARACULO	11	7	7,55	8,02	93%	87%
LUANDA 1	208	9	10,49	13,48	86%	67%
SERRA DA CHELA	103	14	17,71	19,94	79%	70%
LUANDA 2	311	6	6	6	100%	100%
BENGUELA	208	16	18,49	20,15	87%	79%
CELES	104	9	10,49	13,46	86%	67%
MATALA	58	7	8,47	11,42	83%	61%
ONDJIVA	150	10	10,99	12,24	91%	82%
HUAMBO	523	22	39,97	62,42	55%	35%
MALANJE	57	13	17,91	25,28	73%	51%
CUITO	48	7	8,47	11,41	83%	61%

The geographical distribution of the surveyed grid cells in Angola with reliable inventories of bee genera are represented in Figure 2.12, where the well-surveyed grid cells are highlighted.

It is important to take into consideration that richness estimators have a limited performance on samples that are not close to completeness (Gotelli & Colwell, 2001), which is usually the case for terrestrial arthropods in the tropical zones where greater sampling efforts are necessary to capture rare species (Novotny´ & Basset, 2000). Therefore, it is expected that generic richness is underestimated, also due to the non-random nature of sampling present in museum specimens (Petersen & Meier, 2003).

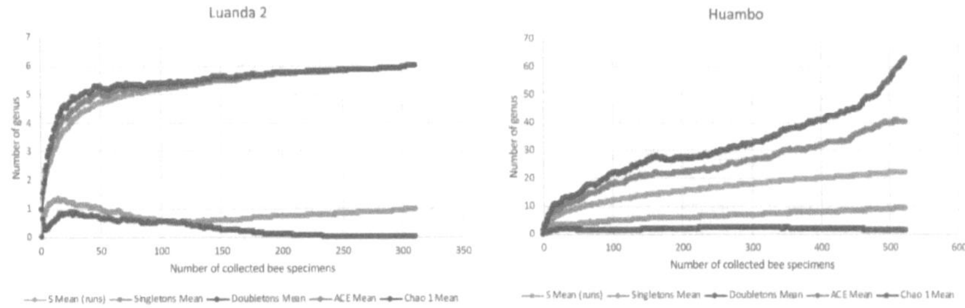

Figure 2.11 – Genera accumulation curve for bee specimens recorded for the Luanda 2 and Huambo half-degree grid cells. Analyses using CHAO 1 and ACE estimates.

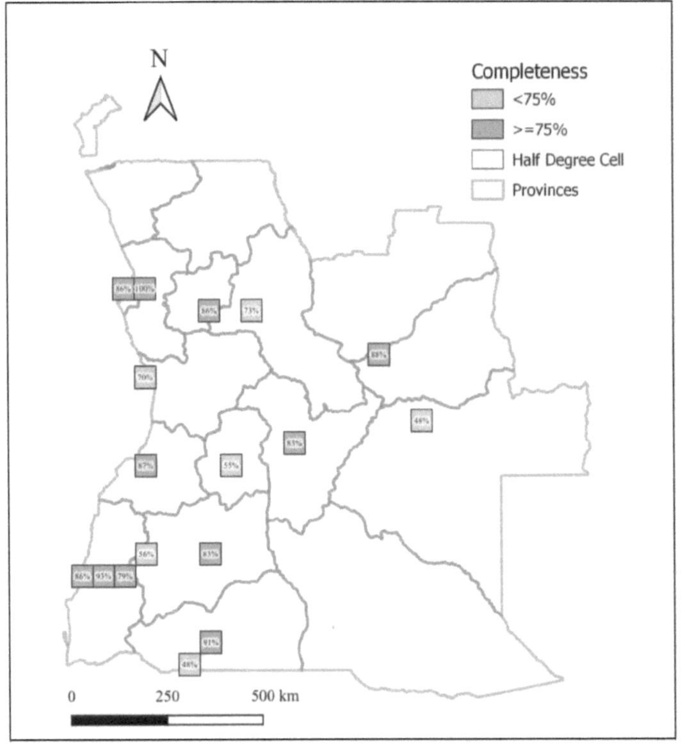

Figure 2.12 – Geographical distribution of the surveyed grid-cells with reliable inventories for bees (Hymenoptera: Anthophila) in Angola. The well-surveyed grid-cells are defined as those with completeness values ≥75% of the value predicted by the estimator (see text above). The map depicts the results for half-degree cell resolution, which corresponds approximately to UTM cells of 25,000km².

From the aggregation of data, as represented in Figure 2.13, it is also clear that December holds the highest number of collections (532; 25%), attributable either to genuine seasonal abundance fluctuations or bias in seasonal collector effort.

To my knowledge, this work is the first contribution to profiling the bee fauna of Angola since the independence of the country, more than forty years ago. The aims of this chapter were to address the knowledge gaps on bees (Anthophila) of Angola, and to integrate all readily available data to provide a foundation profile for this fauna.

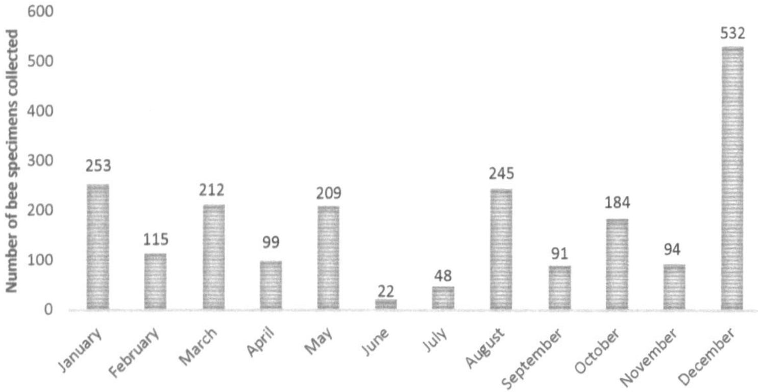

Figure 2.13 - Graphic representation of Angolan bee specimens collected per month, combining data from natural history collections, available online and digitized from the Agronomic Research Institute collections.

The current preliminary checklist of 209 bee species recorded for the country, from 5 families and 47 genera, represents an increase of 82 species when compared to existing profiles. A number of factors suggest that the number of bee species in Angola could be considerably higher, namely: 1) the country has been poorly sampled, with almost half the country (extreme north-west, north-east and east) not surveyed at all; 2) the country has a diversity of habitat types; 3) the variety of climatic features, ranging from tropical humid in the north to extremely arid in the south west would be associated with high species turnover; 4) Angola falls almost entirely into the Zambezian region where levels of species turnover are known to be high; and 5) the country is part of the Southern Africa region where 2,600 species are described, representing about 50% of all Afrotropical bee species (Eardley & Urban, 2010; Kuhlmann, 2009). Therefore, the apparent low diversity of bee species could be an artefact resulting from the disparity between the large size of the country and the lack of bee taxonomists, as well as little research on bee diversity of the region. Additionally, a large proportion of these studies were completed before the independence of the country in 1974, which means the country barely has contemporary data on bee species diversity and abundance.

The studies performed up to the present were also strongly seasonally biased, with most of collections made during the rainy season (December-March). Although seasonal rhythms in tropical communities of insects (e.g. insect assemblage structure) are well established and are related to the seasonal variation in precipitation, with a usual increase in abundance and diversity during the rainy season and a decrease in the dry season (Valtonen, 2013), it would be interesting to determine the seasonality in species composition of the bee fauna. It is recommended that future studies developed in Angola on bee diversity consider the seasonal variation within bee species assemblages, and therefore surveys need to be conducted across all seasons.

The collecting effort with respect to bee fauna has been assessed and reveals a clear collector bias, either around big cities or in specific collecting locations, such as the agronomic experimental stations or along the main roads. A great proportion ($\approx$ 50%) of the country's area remains to be surveyed, lacking representation of significant habitats as the Congolian forest savanna mosaics in the northwest and the Zambezian savannas, woodlands and grasslands in the east, as well as the transition zone between these. This complex of vegetation types, especially in the eastern region of the country, are of immense biodiversity value with some of the most interesting ecosystems in southern Africa, where the vegetation communities occupy deep sands with perched water tables (Werger & Coetzee, 1978), and are lightly populated, which means that habitats remain virtually intact and in pristine condition (Golder Associates, 2006).

Generic richness was found to be strongly associated with collector effort, possibly representing an artefact of the data availability (number of records per grid cell). The geographical distribution of richness most probably better represents intensity of collector effort rather than the richness of species itself. The number of genera currently digitized from natural history represents only 56,1% (23) of the genera expected for the country (41) based on our bibliographic revision, therefore, it is likely that the number of unique genera will increase as more data become available.

Overall, the knowledge on Angolan bee diversity is limited, outdated and strongly biased. Considering the importance of bees as pollinators and thus ecosystem service providers, long-term monitoring studies on their diversity and abundance should be implemented in Angola.

The decades of civil war in Angola contributed for the local extinction of mammals, even in protected areas, as well as for major losses of habitat (Ron, 2015). Promoting the conservation of bee populations will contribute to the rehabilitation of habitats, since reproduction of major elements of the flora is dependent on bees (Michener, 2007; Ollerton *et al*, 2011).

CHAPTER 3. BEE DIVERSITY ALONG A TROPICAL ALTITUDINAL GRADIENT

3.1. INTRODUCTION

Altitudinal gradients represent rapid environmental changes through reduced horizontal distances (over km), with minimal variability in other environmental parameters (Hodkinson, 2005), and have served as a heuristic tool to develop biogeography, ecology and evolutionary biology (Lomolino, 2001). Since the 18th century, when European naturalists started exploring the world, elevational gradients have been considered as optimal models to study responses of species or communities to variations in environmental conditions (Korner, 2007). Two categories of environmental changes are relevant to altitudinal studies: i) general geophysical phenomena, such as changes in atmospheric pressure, temperature and ii) the less altitudinal-specific changes such as changes in moisture, period of sunlight, wind intensity, length of season and even phenomena as fire or human land use (Korner, 2007).

Biodiversity distribution along altitudinal gradients has been mainly characterized by three patterns: i) monotonic decline of biodiversity with altitude (declining); ii) a low plateau where species richness is high until the mid-altitudes before declining (low plateau); and iii) a unimodal distribution with a mid-elevational hump (midpeak) (Grytnes & MacCain, 2007). The difference in patterns is related to the sensitivity of the studies to effects of area, sampling regime, sampling effort (Rahbek, 1995) and scale (Rahbek, 2005).

In tropical mountains with heterogenous habitats, diversity patterns have been mostly illustrated by a generalized decay in species richness with increasing elevations (Morris *et al.*, 2015; Costa *et al.*, 2015; Nunes *et al.*, 2016), with several attempts being made to explain this pattern and its underlying drivers. The following factors have all been considered as driving forces influencing a linear decline of species richness along elevation gradients: increasing rainfall (Devoto *et al.*, 2005), competition between species (Hoiss *et al.*, 2012), climate and habitat management (Hoiss *et al.*, 2013), longer development time (Kocher *et al.*, 2014), habitat preference and behaviour (Koski & Ashman, 2015), reduction in habitat area and resource diversity (Fernandes *et al.*, 2016), temperature (Peters *et al.,* 2016) as well as other physical phenomena such as water vapour pressure and wind speed (Perillo *et al.,* 2017).

Different patterns describing species richness along elevational gradients have also been defined, including: i) a hump-shaped peak of species diversity in midway elevations caused by mid-domain effect, found in long term insect sampling studies (Brehm *et al.,* 2007); and ii) increasing species richness with increasing altitude, found for galling insects in Brazilian high-altitude grasslands (Coelho *et al.,* 2017) and parasitoid wasps in the neotropical region (Veijalainen *et al.,* 2014). However, there is no consensus on the determining factors of biodiversity variation patterns along altitudinal gradients, although a fundamental topic in ecology studies (Hodkinson, 2005; Peters *et al.,* 2016).

Altitudinal gradients have also been used as natural experiments to analyse the influence exerted by changes in climatic conditions on animal communities (Karunaratne & Edirisinghe, 2008). Many species are currently threatened by climate change (Thomas *et al.,* 2004). Climatic warming has already induced alterations in life cycles as well as changes in distribution range, with expansion in the direction of the poles or higher elevations where temperatures are lower (Chen *et al.,* 2011). Climate alterations have impacted biomes and ecosystems, forcing species to either move to more favourable thermal conditions or adapt to the variations in environmental conditions, through behavioural or physiological flexibility (Sinervo *et al.,* 2010; Sunday *et al.,* 2012; Kerr *et al.,* 2015). The failure to adjust or adapt, results in population collapse or even extinction (Sinervo *et al.,* 2010). Species reactions to climate change also depend on life-history traits including tolerance to temperature (heat or cold) (Araújo *et al.,* 2013), and could reveal the common evolutionary history of taxa where lineages were subject to similar historical climatic conditions in the course of their evolution (Kellerman *et al.,* 2012).

Nevertheless, distributional changes seem to be caused by a combination of factors, not just temperature, and it is critical to understand the relative contribution of each factor in order to predict distribution shifts that affect the sustainability of populations or communities (Sunday *et al.,* 2012). Climate change can act synergistically with other stressors of anthropogenic origin, such as land use intensification or usage of agrochemicals, to change species' response to new environmental conditions, causing a decrease in fitness and persistence of populations in tropical and temperate climates (Kingsolver *et al.*, 2011) and induce alterations or even loss of ecosystem services provided by the impacted species (Goulson *et al.*, 2015).

Insects, including non-domesticated bees (Hymenoptera: Anthophila), are exceptional model organisms to assess the effect of climate variation on species richness along altitudinal gradients. Climate variables and weather conditions are of extreme importance for insects (Abrahamczyk *et al.*, 2011), particularly since their capacity for body temperature regulation is determined largely by the ambient temperature (Hozumi *et al.*, 2010). Additionally, bee diversity has proven to be positively related to the availability of food (Araújo *et al.*, 2006) and host plants (Karunaratne & Edirisinghe, 2008), determining factors for their presence or absence at higher altitudes where both trophic resources may be limited (Hoiss *et al.*, 2012).

Therefore, community composition of bee species is expected to decrease with increasing altitude (Hoiss *et al.*, 2012). Furthermore, under a scenario of mean global temperature increases, both the diversity and abundance of wild bees will be affected (Widhiono *et al.*, 2017), because the foraging activity, body size and lifespan are all influenced by high temperatures, which in turn affects patterns of pollen flow and overall pollination success (Scaven & Rafferty, 2013). Likewise, as floral resource availability is modified by warming, this will impact the fecundity of pollinating insects (Scaven & Rafferty, 2013). The overall result could be extensive trophic disruption (Scaven & Rafferty, 2013).

The study of wild bee diversity at various altitudes in the tropics could provide insight on possible responses of bee communities to climate variations, and how the ecosystem services provided by these communities would be altered (Scaven & Rafferty, 2013).The southern African Great Escarpment, a 5000 km-long semi-continuous mountain chain, ranges from north-western Angola to eastern Mozambique, offering the ideal environment for altitudinal gradient research studies. These mountain systems are expected to host half of the subcontinent's centres of plant endemism and a rich endemic vertebrate fauna and provide a major proportion of the subcontinent's fresh water (Clark *et al.*, 2011). To ensure the continuous provision of these ecological services, this montane habitat and its rich biodiversity urgently need to be protected and restored (Clark *et al.*, 2011). Climatic conditions change along altitudinal gradients, therefore biodiversity surveys along such gradients allow predictions of the possible impact of climatic change on biodiversity patterns. These predictions are based on the assumption that the climate changes along gradients in the same way it changes in time, known as space-for-time substitution approach (Dunne *et al.*, 2004). Intensive biodiversity systematic surveys to assess the effects of climate change on the African Escarpment will provide vital information to develop effective environmental management (Clark *et al.*, 2011).

Even though the general knowledge on the escarpment's biodiversity is poor, fragments of it have been studied in some detail (see van Wyk & Smith, 2001 for examples), but Angola remains the least-known section (Figueiredo *et al.*, 2009). The Angolan Escarpment, a nearly 1000 km-long mountain range, supports a gradient of vegetation types from tropical evergreen and semi-deciduous rainforest in the north, Afromontane forest-grassland mosaics and miombo woodland in the centre, to Kalahari Highveld shrublands and Nama-Karoo semi-desert in the south (WWF, McGinley, 2008). Furthermore, it supports high levels of vertebrate endemism, surpassed only by South Africa (Clark *et al.*, 2011). In particular, the 'Angolan Escarpment Woodlands' (Olson & Dinerstein, 1998), a narrow band of montane woodlands found between 350 and 1000m above sea level, can be considered to be a 'biodiversity hotspot' (*sensu* Myers *et al.*, 2000) due to the high levels of critically endangered biodiversity (Clark *et al.*, 2011).

The decades of civil war in Angola halted all ecological research but recently studies have resumed in areas of high endemism or conservation concern in the Angolan escarpment (Lovett, 2018). A few botanical studies (Hind & Goyder, 2016; Gonçalves & Goyder, 2016; Gonçalves *et al.*, 2016), primate research (Bersacola *et al.*, 2015; Svensson *et al.*, 2017), herpetological surveys (Baptista *et al.*, 2018) and assessments of deforestation, degradation and carbon stocks (Leite *et al.*, 2018) have been conducted along the escarpment, mostly concentrated in the central region. However, the majority of the research in the Angolan Escarpment has been dedicated to ornithology (Ryan *et al.*, 2004; Sekercioglu & Riley, 2005; Mills & Dowd, 2007; Mills & Dean, 2007; Mills, 2010; Mills *et al.*, 2012; Cáceres *et al.*, 2014; Cáceres *et al.*, 2016; Cáceres *et al.*, 2017; Baptista & Mills, 2018). Nevertheless, the entire area (and particularly the southwestern region) remains poorly documented for many taxa, especially insects.

To my best knowledge, the present study is the first to assess the species richness and composition of the most important pollinating insects - bees (Hymenoptera: Anthophila) along a tropical altitudinal gradient of the southwestern Angolan Escarpment.

I explored the following hypotheses:

1) Whether taxon richness decreases with increasing elevation along the tropical altitudinal gradient, as has been reported for both temperate and tropical altitudinal studies elsewhere (Arroyo *et al.*, 1982; Perillo *et al.*, 2017)

2) Whether community composition changes with increasing altitude as has been demonstrated on studies in temperate (Hoiss *et al.*, 2012) and tropical zones (Classen *et al.*, 2017), as bee community structure is also known to be affected by landscape characteristics at different scales (Tscharntke *et al.*, 2012).

3.2. METHODOLOGY

3.2.1. STUDY AREA & SITES

The fieldwork was undertaken in the Angolan southwestern escarpment, situated close to Lubango, the capital of Huíla Province, in the valley of Bruco, a southeast-northwest transect stretching between the higher (S15.15; E13.26, WGS84) and lower (S15.12; E13.19, WGS84) altitudes of Serra da Chela (Figure 3.1). Serra da Chela is an imponent mountain massif composed of a series of geomorphological plateaus that are defined by escarpments of up to 1000m slopes down to the inferior level (Diniz, 2006). The Humpata Plateau (2000 m.a.s.l.) is one of the highest plateaus of Serra da Chela, and Bruco is located on its western edge. The altitudinal span of the transect was 760-1651 m.a.s.l.

The escarpment of Serra da Chela is comprised of quartzitic blocks that form a cornice over a sedimentary zone, of Cambrian-Silurian age, composed by layers of schist and dolomitic limestones that overlay antechamber eruptive formations, mostly of granite (Diniz, 2006). Materials from the Plio-Pleistocene can be found as gravel along the main rivers and at the lower altitude of the valley, as terraces that follow the river or as sloped deposits such as pebbles and blocks of conglomeratic rocks (Diniz, 2006). Deposits from the Holocene are more restricted and can be found in the bottom of secondary rivers or on the borders of main rivers flowing down the slope (Diniz, 2006).

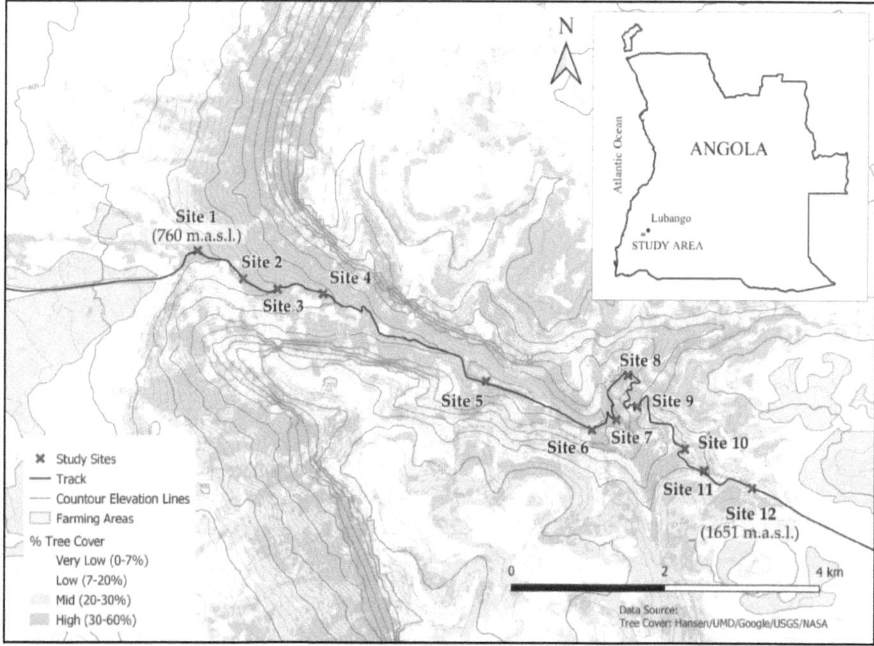

Figure 3.1 – Topographical map showing the percentage of tree cover for the transect within the Bruco valley with location of the twelve sampling sites and indication of lowest and highest altitude. The location of the transect area is shown within Angola and using the capital of Huíla province – Lubango – as reference (insert).

Pedologically, the study area is mainly composed of lithosols associated with rocky outcrops or terrain, displaying an increase in the frequency of rocky elements and reduction of soil thickness with increasing altitude (Diniz, 2006). The soils in the study area are of poor agricultural potential, although along the margins of the river at the lower flattens the deposition materials are very heterogenous with higher mineral reserves and therefore more fertile than the rest (Diniz, 2006).

The study area represents a transition between two biogeographical regions: Central plateau and Namibe (Rodrigues *et al.,* 2015). The Central Plateau unit is integrated in the Zambezian region (Linder *et al.,* 2012) and includes a great proportion of the WWF Ecoregion of the Angolan miombo woodlands (Olson *et al.,* 2001). The Namibe unit is part of the south-western Angola region (Linder *et al.,* 2012), which corresponds with the WWF Ecoregion of the Kaokoveld desert, representing the northern part of the Namib desert and Namibian savanna woodlands (Rodrigues *et al.,* 2015).

The vegetation on the plateau that extends to the higher altitude of the valley consists of Miombo woodlands, which open to dense savannas, dominated by *Brachystegia* species (van Jaarsveld, 2010). In this area the vegetation is shaped by grass fires, mainly occurring in the dry season, and sandy soils with poor mineral content (Rodrigues *et al.,* 2015).

Figure 3.2 – Landscape of the Afromontane forest and valley escarpment of the study area.

Along the gorges (Figure 3.2), the vegetation changes into moist, deciduous broadleaf woodlands of Afromontane Forest with species such as: *Ficus bubu, F. sur, F. burkei, F. natalensis, Sclerocarya birrea, Mangifera indica, Aloe vallaris, Plectranthus cylindraceus, Senecio orbicularis, Haplocarpa lyrata, Gloriosa superba, Aeollanthus rehmannii, Kalanchoe lanceolata, K.laciniata, Crassula expansa fragilis, Scadoxus multiflorus, Carissa spinulifera, Solenostemon rotundifolius, Adansonia digitata, Podocarpus milanjianus, Buxus macowanii, Trema orientalis, , Doryopteris concolor, Pteris sp., Pleopeltis sp., Obetia carruthersiana, Euphorbia vallaris,* among others (van Jaarsveld, 2010).

Close by to the lower altitude of the transect, the vegetation transforms into dry bushveld dominated by *Pachypodium lealii, Adansonia digitata, Colophospermum mopane, Terminalia prunoides, Croton* sp.*, Aloe littoralis, Ximenia caffra, Balanites welwitschii, Cordia* sp.*, Euphorbia subsalsa, Plectranthus amboinicus,* and *Sterculia africana* to name a few (van Jaarsveld, 2010). The Namib Desert lies not far (80km in straight line) to the southwest. The transect thus runs from a semi-arid zone through to subtropical Afromontane forest.

Due to the shortage of suitable browsing conditions in the dry season, the reduced quality of leaf matter and inaccessibility of tree canopies, the area on the higher altitude and along the valley have relatively low density of mammals and are occupied by a few widespread species such as blue duiker (*Philantomba monticola*), Malbrouck monkey (*Chlorocebus cynosuros*), blue monkey (*Cercopithecus mitis*), Chacma baboon (*Papio ursinus*), African civet (*Civettictis civetta*), serval (*Leptailurus serval*), caracal (*Caracal caracal*), leopard (*Panthera pardus*), as well as genets and moongooses (Huntley, 1974; *pers. obs.*).

3.2.2. SITES & SAMPLING DESIGN

- **Study Sites**

In January 2017, 12 selected sites along an elevation gradient in the Bruco valley (Serra da Chela, Angola) were used to survey for bee species richness and community change (Figure 3.1).
Species richness and abundances were determined for each altitude sampled enabling the determination of bee species assemblages and, consequently, the analysis of possible community change along the elevation gradient.

The sites were selected with an altitude range from ca. 760 to ca. 1651 m.a.s.l. and the selection criteria were: 1) location along the altitudinal gradient, with 60-80 m altitude difference between sites; 2) landscape accessibility (some areas were extremely rocky and with coarse gravel, therefore not suitable to sample); and 3) permission from local population to access the area. The ca. 11 km transect was largely inaccessible by car (apart from a few lower sites), and included very steep gradients.

- Data collection

Bees (Hymenoptera: Anthophila) were collected from January 5th to January 24th in 2017. On all 12 sites, bees were sampled between 8 a.m. and 4 p.m. on sunny days, when temperature ranged from 18°C to 24°C. Following Jiménez-Valverde & Lobo (2005), a combination of sampling methods was applied to achieve a reliable inventory of bee species, viz.: i) sweep netting; ii) pan traps; and iii) malaise traps.

At each site, one malaise trap was set up (Figure 3.3-a) in the early morning and recovered in the afternoon due to security reasons (theft). The technique proved to be inefficient due to destruction of the traps by grazing cattle, and consequently had to be discontinued.

Replicates of colour pan traps were also placed on all 12 sites (Figure 3.3-b), with 5 sets of 3 different coloured pan traps (white, blue, yellow; 15 cm diameter containers filled with water and a drop of liquid soap to break the surface tension of the water; N=180) used on each site, placed on the ground with a distance of 20 m between sets of traps. Pan traps were kept for 72 hours per sampling period and specimens were recovered, every afternoon to avoid losses. These were transferred to plastic vials and preserved in 96% ethanol, collecting a total of 164 bee specimens. During field work, a considerable number of pan traps were stolen, but it was possible to maintain a minimum of 3 replicates per site.

Sweep netting and directed netting proved to be the most efficient sampling techniques, allowing the collection of 685 specimens along the 12 sampling sites. For each site, sampling followed randomised transect walks along the viable accesses (Figure 3.3-c). The opening of the sweep net was 40 cm in diameter and two sweep net approaches were applied: i) transect walks were subdivided into 10 assemblages of 50 sweeps each, with the net being emptied at the end of each of the 10 sub-transects; and ii) transect walks where all bee specimens observed were collected over a 60 minute period, with the net being emptied at regular intervals to avoid loss or destruction of bee specimens.

Figure 3.3 – Sampling methods used at each sampling site: a) malaise trap; b) five sets of three pan traps; and c) sweep netting.

Bees were identified to the lowest taxonomic level possible based on Michener (2007), assisted by comparison with the bee collection at the Entomology department, Iziko museums. These identifications were later ratified by a bee taxonomist with expertise in southern African bee taxonomy (C. Eardley). For some poorly-studied genera, it was not possible to distinguish species or even morphospecies and for that reason the data analysis was largely performed at the genera level. Specimens were deposited and accessioned in the Entomology Collection in Iziko Museums of South Africa, Cape Town.

Environmental variables representing factors known to influence altitude-diversity relationships, such as annual precipitation (mm) and mean annual temperature (°C), were obtained for each site using Worldclim version 1.4 (1960-1990). I then tested for effects of these environmental variables on generic richness and composition along the altitudinal transect by regressing generic richness and diversity with altitude (Appendix III).

Bee community structure is also known to be affected by landscape characteristics at different scales (Tscharntke *et al.*, 2012). For that reason, I tested the effects of landscape variables such as Normalized Difference Vegetation Index (NDVI) and tree canopy cover on generic richness, at two spatial resolutions of 200 and 400 m radius centred on each sampling site. This radius was chosen based on the fact that most of the bee species present in the study area were small solitary bees (Chapter 4), known to have foraging ranges smaller than 600m (Greenleaf *et al.*, 2007). NDVI was computed at the patch (200 m radii) and landscape scale (400 m radii) using bands 4 and 8 from Sentinel 2+image, acquired on 13 January 2017. Tree canopy cover for year 2000 was obtained from Hansen *et al.* (2013) and is defined as canopy closure for all vegetation taller than 5m in height, encoded as a percentage per output grid cell, in the range 0–100. The landscape analysis was performed using QGIS 2.4.

3.2.3. STATISTICAL ANALYSES

- Sampling completeness

In order to evaluate the completeness of the genera inventory sampled, I calculated the cumulative number of genera found at each sampling site and for the entire altitudinal transect, using species accumulation curves in the software EstimateS (Colwell, 2013; version MAC 9.1.0). The data obtained from the sweeping transects and pan trapping were used as replicates for generic richness estimations. The asymptotic value of the accumulation curves was used together with the generic richness estimation to test if the total number of genera caught in each sampling plot, and along the entire transect underestimated the true generic richness. The non-parametric richness estimators computed were Chao1, Chao2, incidence-based coverage estimator (ICE), abundance-based coverage estimator (ACE) and first and second order Jackknife. Detailed descriptions of the estimators can be found in Gotelli & Chao (2013). To derive confidence intervals, one hundred randomizations of sampling order (without replacement) were carried out for each site and for the entire transect. According to Hortal *et al.* (2006), the estimators ICE, Chao

2 and Jackknife1 and 2 perform well under low sampling intensities, presenting reduced sensitivity to sample coverage, unevenness of species distributions or variability in capture probability. In line with other studies on bee diversity along altitudinal gradients in the tropics (e.g. Perillo *et al.*, 2017), the estimator Jackknife1 was used to calculate sample completeness.

- Generic diversity along the altitudinal gradient

The data obtained during field work were used to calculate three measures of diversity, two emphasizing diversity and one equitability/evenness: i) the Shannon diversity index (H'); ii) the Margalef's index (d); and iii) the Pielou's evenness index.

The *Shannon diversity index* (H') is obtained through the equation

$$H' = -\sum_i p_i \, log_e \, (p_i)$$

where p_i is the proportion of bee genera of the ith genera in the entire collection (Clarke & Warwick, 2001).

The *Margalef's index* (d) is used as an alternative to generic richness. Generic richness (S) being simply the total number of genera, is obviously highly dependent on sample size (Clarke & Warwick, 2001). The Margalef's index incorporates the total number of individuals (N) and measures the number of genera for a given number of individuals, following the equation:

$$d = \frac{(S - 1)}{\log N}$$

The *Pielou's evenness index* (J') expresses the extent to which individuals are evenly distributed among the different genera present in the community and can be obtained through the equation:

$$J' = \frac{H'}{H'_{max}} = \frac{H'}{logS}$$

where H'_{max} refers to the maximum Shannon diversity value, obtained if all species were equally abundant ($logS$) (Clarke & Warwick, 2001).

All diversity measures were computed using PRIMER-E version 6.1.11. Correlation matrices and t-tests were performed using Statistica version 13.5.

- Community analyses

To examine the bee communities along the altitudinal gradient I first generated a Bray-Curtis sample dissimilarity matrix of square root transformed generic abundances for data pooled at the site level. To evaluate community similarity an analysis of similarity (ANOSIM) was applied. The hierarchical cluster analysis allowed the definition of distinct community relationships among sites, represented in a dendrogram. To identify the genera that most contributed to the dissimilarity between groups I applied a similarity percentages (SIMPER) analysis. In order to evaluate the inter-relations between group samples, a non-metric multidimensional analyses (NMDS) was performed using a two-dimensional ordination. Sites similar in structure were plotted in close proximity in the ordination space, while sites distantly plotted will be less similar. The goodness of fit of the ordination is defined by the stress index that should present a low value (at most <0,1; Clarke & Warwick, 2001) in order to provide a realistic approximation of the true relationship between sites.

Geographical (altitude), environmental (annual mean temperature and annual precipitation) and landscape variables (tree cover at 200 and 400 m radii and NDVI at 200 and 400 m radii) were normalised prior to the analyses and used for principal coordinates analysis (PCO) based on Euclidean distance. The PCO analysis was also performed in PRIMER-E version 6.1.11 to investigate interpoint dissimilarities of bee genera and variables such as the geographical, environmental and landscape parameters. All community analyses were performed.

3.3. RESULTS & DISCUSSION

3.3.1. Sampling summary

On the 12 sampling sites, a total of 845 individuals were collected, representing 51 identified species, 34 genera and five bee families (see Appendix IV, Table 3.1). The dominant families at all study sites were Apidae (62%), Megachilidae (20,6%) and Halictidae (16,6%). Specimens of Colletidae were only recorded at two sites. Only one specimen of Andrenidae and Melittidae were collected, interestingly at the highest altitude sites. Among the Apidae species, the most abundant were *Hypotrigona ruspolii*, followed by *Macrogalea candida* and *Braunsapis facialis*.

Even though the diversity analysis performed in this chapter is at the genera level due to the limited number of individuals identified to the species level, it is worth to mention that this study contributed to the knowledge of wild bees of Angola by adding 23 species and 6 genera as new records for the country's preliminary checklist presented on Appendix II. The preliminary checklist of bees of Angola is now composed by 209 species, from 47 genera and five families.

Table 3.2 – Number of individuals and genera collected, as well as sample completeness as calculated with Jackknife 1 estimator, per sampling site.

Data properties			Estimator values (SD)						Coverage (%)
Altitude m.a.s.l.	Sobs	N	Chao1	Chao2	ICE	ACE	Jackknife1	Jacknife2	Jackknife1
760	17	84	22,19 (5,32)	44,73 (26,3)	36,6	24,82	27,08 (6,27)	34,73	62,78
840	12	107	12,2 (0,62)	12,39 (0,86)	14,61	13,12	14,75 (1,44)	12,7	81,36
901	11	84	12,48 (2,57)	15,13 (4,44)	27,06	13,33	16,5 (3,18)	18,47	66,67
960	11	43	21,25 (10,34)	48,13 (45,03)	51,2	23,03	19,25 (4,09)	25,99	57,14
1120	17	77	18,48 (2,21)	22,13 (4,95)	23,87	19,97	24,33 (5,64)	27,97	69,87
1179	18	72	23,18 (5,31)	25,64 (6,38)	28	26,77	27,17 (5,56)	30,95	66,25
1253	20	117	26,94 (6,6)	29,17 (7,72)	35,59	30,12	29,17 (6,47)	33,71	68,56
1305	11	40	12,95 (2,82)	13,29 (2,93)	15,61	14,28	15,58 (3,7)	17,48	70,6
1489	11	37	12,95 (2,82)	13,29 (2,93)	15,61	14,28	15,58 (3,7)	17,48	70,6
1534	17	97	30,86 (13,12)	67,42 (32)	41,66	27,86	27,08 (6,42)	36,25	62,78
1580	18	57	32,74 (12,64)	36,49 (16,02)	35,95	34,74	28,08 (5,49)	34,98	64,1
1651	11	30	11,97 (1,77)	12,1 (1,71)	13,92	12,35	14,67 (2,82)	14,97	74,98
Entire transect	35	845	37,5 (3,16)	36,31 (1,8)	37,88	38,36	39,58 (2,12)	39,2	**88,43**

A smoothed species accumulation curve was computed for each study site (see Appendix V) and none reach an asymptote, although sampling coverage across sites ranged between 57,14% and 81,36% (Table 3.2). The species accumulation curve for the entire dataset (i.e. all sites pooled) appeared to level off indicating the sampling of a large proportion of the genera pool.

According to the Jackknife1 estimator, sampling completeness for the entire transect was 88,43% (observed richness: 35; estimated richness: 39,58), as summarized in Table 3.2. The sampling recorded a low number of rare genera, with 5 singletons (14,29% of the total) and 3 doubletons (8,58% of the total).

Single diversity measures were calculated for each site pooling data from the total sampling because some of the replicates presented a limited number of specimens (Table 3.3).

Table 3.3 – Diversity measures (Margalef [d]; Shannon indexes [H']; and (Pielou's [J'])) for bee genera at the 12 sampling sites along the altitudinal gradient, with 95% confidence intervals. S = number of species; N = abundance.

Altitude	S	N	d	J'	H'(loge)
760	17	84	3,611	0,756	2,143
840	12	107	2,354	0,686	1,704
901	11	84	2,257	0,681	1,632
960	12	43	2,925	0,606	1,506
1120	18	77	3,914	0,881	2,546
1179	18	72	3,975	0,856	2,475
1253	21	117	4,200	0,785	2,389
1305	11	40	2,711	0,836	2,005
1489	11	37	2,769	0,865	2,075
1534	17	97	3,497	0,774	2,194
1580	18	57	4,205	0,822	2,376
1651	11	30	2,940	0,942	2,258

It was not possible to detect a conspicuous pattern in the generic diversity along the altitudinal gradient, though the analysis reveals a discrete increase in diversity with increasing altitude, a pattern that is contrary to what has been detected along altitudinal gradients of temperate (Arroyo *et al.*, 1982) and tropical ecosystems (Perillo *et al.*, 2017). Bee community diversity was highest (*H'* = 2.546) at a mid-transect site (altitude = 1120 m.a.s.l.). This might be an artefact related to the fact that this sampling site had a cropped area at the forest fringe which is in accordance to other studies that found bee diversity to be higher in low agricultural intensity areas surrounded by high proportions of semi-natural habitats (Basu *et al.*, 2016).

Studies on animal diversity patterns along gradients (elevational or latitudinal) most frequently explain the patterns using the environmental variables. For instance, cold temperatures found in high elevations are expected to be physiologically tolerated only by a limited number of species; the speciation rates can also be influenced by reduced generation times or increased metabolic rates (Gillman *et al.*, 2012); or even, species diversity might be maintained by acceleration of density-dependent mortality with temperature (Bagchi *et al.*, 2014).

The regression between the univariate measures and altitude or the environmental variables known to affect bee diversity (annual mean temperature and annual precipitation) revealed positive relationships between evenness (J') and altitude (r = 0.68, p < .001) and annual precipitation (r = 0.60, p < .001), with a negative relationship with annual mean temperature (r = -0.66, p = .015) (temperature decreased with altitude). The relationships of the other univariate measures with the same variables proved to be statistically marginal as summarized in Table 3.4 (see Appendix VI for charts).

Nevertheless, NDVI is also frequently used as a representation of energy or resources (Hawkins *et al.*, 2003) that could affect the patterns of species diversity along altitudinal gradients. In fact, for endothermic organisms (e.g. birds), a strong positive relationship has been found between species richness and resources availability (Price *et al.*, 2014). However, for ectothermic insects the usage of resources availability is determined by ambient temperature and the energy consumption during foraging declines with decreasing temperatures, an event that is overcome by foraging only above a certain temperature threshold (Stabentheiner *et al.*, 2003). The limitations of resource exploitation as determined by temperature could then reduce population densities in cooler temperatures and increase extinction probability (Classen *et al.*, 2015).

The result of testing the univariate measures of diversity with the landscape variables along the altitudinal transect, revealed a significant negative relationship between evenness (r = -0.71, p < 0.001) and diversity (r = -0.63, p < 0.001) with tree cover at the patch scale (200m radii). However, at the landscape scale (400m radii), a significant positive relationship was found between diversity (r= 0.63, p < 0.001) and the NDVI. The bee abundance was not strongly affected by any of the landscape variables at both scales.

The NDVI reflects the net primary productivity at small spatial scales (Wu *et al.*, 2014) and high levels of vegetation cover can reduce the survey area, sample efficiency or shift pollinator activity to canopy level, negatively influencing the sampled diversity (Hurlbert, 2004). Though, my results indicate that bee diversity was higher in areas with greater NDVI, this relationship was stronger at the landscape scale. This result could be explained by the fact that the transect used in the study followed mostly a trail that was surrounded by flowering plants and shrubs, an open (disturbed) habitat along the altitude gradient. Open habitat is known to support high bee diversity (Gikungu *et al.*, 2011).

Table 3.4 – R values for the regression between univariate measures and altitude as well as environmental and landscape variables, calculated for the twelve sampling sites: N – abundance; d – Margalef's index; J' – Pielou's index; H' – Shannon diversity index (H').

Variables / Indices	N	d	J'	H'
Altitude	-0,4305	0,2634	0,6827	0,4945
Annual Mean Temperature (°C)	0,3295	-0,3522	-0,6595	-0,5534
Annual Precipitation (mm)	-0,3325	0,2739	0,5957	0,4693
Tree cover (mean; 200m)	0,1619	-0,3605	-0,7122	-0,6324
Tree cover (mean; 400m)	0,1349	-0,1076	-0,4255	-0,3018
NDVI (mean; 200m)	0,0582	0,3958	0,3286	0,4746
NDVI (mean; 400m)	0,0299	0,4897	0,5043	0,6349

However, the abundance of flowering plants or shrubs layers are limited by dense canopy cover which is in line with the results of this study since a negative correlation between bee generic diversity and tree cover was found.

3.3.3. Community analysis

The cluster analysis identified four groups that were >40% dissimilar to one other (Groups A, B, C and D (Figures 3.4 and 3.5).

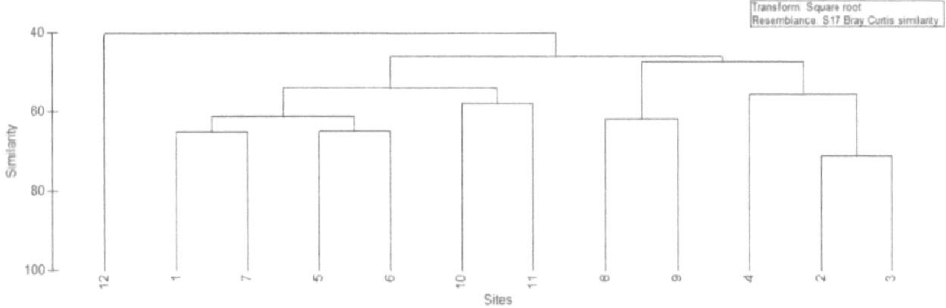

Figure 3.4 – Dendrogram using group-average linking on Bray-Curtis genera similarities from standardized abundance data, transformed with square root. Four groups are defined at similarity level of 50%.

The four groups were clearly outlined on nMDS ordination plot with stress value of 0.11, indicating 'good' 2-dimensional representation of the community structure (Figure 3.4). The overlay of the clusters on the MDS is represent in Figure 3.5. Group B (sites 8 and 9) reflects an assemblage of low altitude sampling sites, as does group C (sites 2, 3 and 4) with medium altitude sites. Group D is composed only by the highest altitude site where bee community was significantly different from all other sites. Group A was composed by a mixture between sites of low, medium and high altitude (sites 1, 5, 6, 7, 10 and 11).

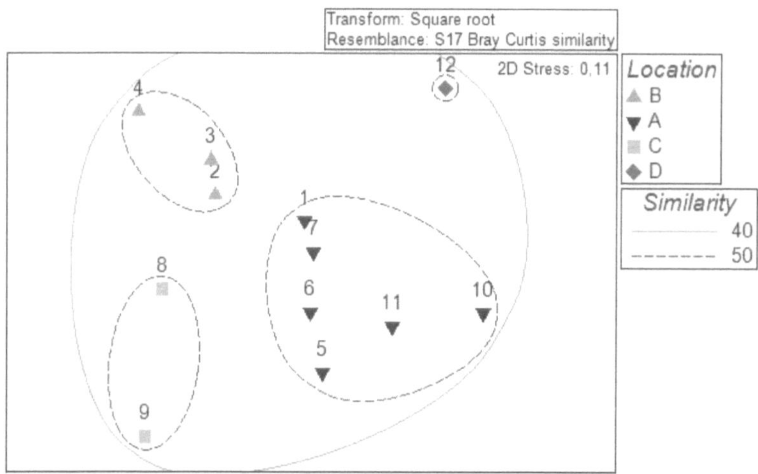

Figure 3.5 – An ordination plot derived from 2-dimensional MDS of assemblage composition of sampling sites based on Bray-Curtis similarity indices. The four groups from the cluster analysis are indicated (stress = 0,11). Numbers are sites, with 12 being the higher altitude of the transect.

The ANOSIM analyses used to validate community similarity clearly reflects statistically significant differences between bee communities of the various groups, presenting a significant level (p =0.1) and a mid-range value of R^2 (= 0.669) for the global test of groups A, B, C and D. Group A differed from B (r= 0.73), C (r=0 .84), and D (r= 0.93). Group B was also significantly different from groups C (r= 0.75) and D (r= 1).

The SIMPER analysis identified genera that defined the various site clusters and % values are individual contributions of genera to observed % dissimilarity between groups. For group A, the genera that most contributed most to this community were *Macrogalea* (13,34%), *Ceratina* (11,79%) and *Braunsapis* (10,40%). In group B, *Hypotrigona* (38,83%) is clearly the main driver for the community assemblage but *Pachyanthidium* (11,26%) and *Ceratina* (10,92%) also contributed.

Group C had the highest average similarity among its sites (61,88%), with *Hypotrigona* (22,77%), *Macrogalea* (14,91%) and *Braunsapis* (12,17%) being the most characteristic genera for these lower sites. The highest average dissimilarity between groups was observed between groups C and D (67,29%). *Hypotrigona* is the genera that most influenced dissimilarity between groups contributing as high as 21,8% to the dissimilarity between groups B and D, and to a lesser extent between groups A and B (14,10%) and B and C (13,67%).

The principal coordinate analysis (PCO) allowed an investigation of inter-point dissimilarities between sites according to geographical, environmental and landscape variables (Figure 3.6). The first two axes of the PCO plot explained 48,8% of the total variation. Defined groups were associated with altitude, precipitation and NDVI variables, both at patch and landscape scales. Group B was mainly influenced by temperature and tree cover at the patch scale and group C was influenced by tree cover at the landscape scale.

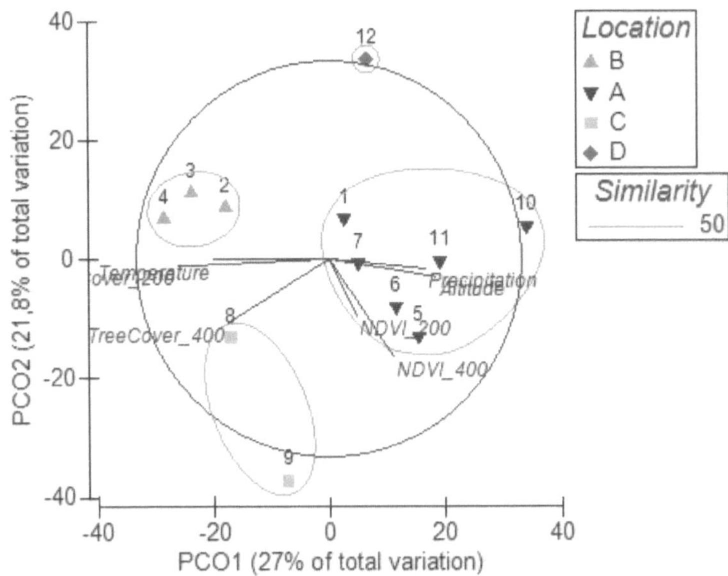

Figure 3.6 – Principal coordinate analyses (PCO) of geographical, environmental and landscape variables based on Euclidean distance and bee generic composition based on Bray-Curtis similarity from square root transformed data. Different colors indicate groups of samples and the green line separates groups.

A potential caveat of this study is that the taxonomic identification was only possible to the generic level. Although species names were obtained for many taxa, difficult and understudied genera could not be identified to species, even by local taxonomists with expertise in southern African bees. This has some advantages when applying multivariate methods of statistical community analysis (Clarke & Warwick, 2001), but also potentially influences the net analysis. It is possible that analysis at a lower taxonomic level (species) would result in a more refined result, with more subtle patterns emerging, particularly as responses to subtle environmental gradients are more likely to be detected in species rather than generic responses.

Overall, bee diversity increased with increasing altitude (in contrast to what has been found by the majority of both temperate and tropical altitudinal gradient studies; Hoiss *et al.*, 2012; Classen *et al.*, 2017; Perillo *et al.*, 2017). How these findings relate to each other is not entirely clear but differences among the range of temperatures felt on the different study areas could be a possible explanation.

For the temperate region (Hoiss *et al.*, 2012), the study was conducted in the German Alps and highest point of the transect was at 2000 m.a.s.l., and although there is no specific indication of the temperature range between sites, it showed that high elevation sites were less sampled due to the shorter period without snow which leads to the thought that temperatures were generally lower than the ones along Bruco transect. For the tropical region, the altitudinal gradient on the Mount Kilimanjaro study (Classen *et al.*, 2017) was considerably longer, reaching 4000 m.a.s.l. at the highest point where mean annual temperature is 0°C. Sampling at Caraça Mountains in Brazil (Perillo *et al.*, 2017) even though it occurred in a similar altitudinal gradient length and range, had lower annual mean temperatures (13 - 19°C) while the lowest annual mean temperature on the Bruco transect was 18°C for the highest site. Annual mean temperatures along the entire Bruco transect (18-24°C) fall within the range of temperatures that favour bee activity, therefore the results obtained in this study could potentially be better explained by a combination of factors (e.g. temperature, rainfall, vegetation communities), as found elsewhere (Classen *et al.*, 2015), rather than just attributed to a single environmental variable.

Globally and contrary to the research hypothesis initially formulated, a clear relationship was found between bee species richness and altitude as the generic diversity of bees along the altitudinal gradient in Bruco pass – Serra da Chela, Angola – increased with increasing elevation.

This result is also contrary to what has been found in other studies, but transect length, temperature ranges and most probably vegetation communities are also not comparable to the ones present in other studies performed in the tropical zones (Perillo *et al.*, 2017; Classen *et al.*, 2017).

For altitudinal transects where temperature at high elevation sites limits bees' foraging capacity, influences generation times and slows evolutionary and diversification rates (e.g. in temperate regions) temperature might act as filtering effect and regulate community assemblage (Hoiss *et al.*, 2012). In tropical regions, the variations in bee community assembly along shorter (Perillo *et al.*, 2017) or longer (Classen *et al.*, 2017) altitudinal transects is possibly better explained by a combination of factors, as it is expected to happen along the Bruco altitudinal transect.

The diversity of bee species (51) and genera (34) found along the Bruco transect is significantly high when compared with the results obtained in similar studies elsewhere in the tropics (41 species in Brazil, Perillo *et al.*, 2017), especially considering that 25% of the samples are yet to be identified to the species level. The fact that this study added 23 new species records for the country is a clear indication of how poorly documented the biodiversity of Angola is. Bee communities did change along altitudinal gradients, although it was not possible to determine whether this was related to temperature, rainfall or ecological factors such as changes in plant communities.

Considering the limitations of this study, additional investigations should be considered to improve the knowledge of how bee communities are influenced by different variables along altitudinal gradients, including studies in both seasons to detect seasonality changes. Land use and habitat disturbance have not been assessed in this study, but both are known to influence bee diversity (Winfree *et al.*, 2009; Potts *et al.*, 2010; Gess & Gess, 2014). Collecting this information will contribute to understand how wild bees respond to climate change and land use, at different scales.

CHAPTER 4. BODY SIZE VARIABILITY OF BEES ALONG A TROPICAL ALTITUDINAL AND RAINFALL GRADIENT

4.1. INTRODUCTION

Life-history traits of species within community assemblages vary along altitudinal gradients, providing an analogous situation to environmental changes that follow global warming (Diamond *et al.,* 2011). Body size is an important life-history trait responsive to natural selection, mainly associated with resource capture, fecundity and mortality (Chown & Gaston, 2010). Furthermore, the differences in dispersal ability associated with body size determines the annual shift in species response to weather conditions as well as the regional variation in species distribution ranges (Stevens *et al.,* 2010).

Patterns in body size variation, both intra- and intraspecific, are of great significance to evolutionary ecology and the most popular generalization is that body size of both endo- and ectothermic organisms increases with altitude (Hodkinson, 2005), a variation of Bergman's rule. When Bergmann first put forward this rule in 1847, he noted that endothermic animal species were smaller in warmer climates and that species living under cooler temperatures had larger sizes (Meiri & Dayan, 2003). Bergmann hypothesized that organisms with larger body mass would lose smaller amounts of energy as their surface-to-body volume ratio was smaller. The rule has since been reformulated, and found to be only applicable to intraspecific variation and ectothermic organisms (Brehm & Fiedler, 2004). Subsequently, it was modified to apply to ectothermic animals through the size-temperature rule, where body size increases with decreasing temperature (Atkinson, 1994). Its application to latitudinal gradients alone (Blackburn *et al.,* 1999) was rapidly refuted and the Bergmann rule has been extended to altitudinal gradients (Smith *et al.,* 2000).

More recently, evidence for increasing body size with altitude has been supported by research on birds and mammals, where the majority of the studied species followed Bergmann's rule (Meiri & Dayan, 2003). However, it has not been demonstrated for freshwater fishes (Fu *et al.,* 2004) and ambiguous results have been obtained for amphibians, which displayed patterns of both increasing (Liao & Lu, 2012) and decreasing (Cvetkovic'*et al.,* 2008) body size with elevation.

Regarding insects, the variation of body size with altitude does not seem to follow a regular pattern among the different insect orders. Insects belonging to certain orders have been shown to increase body size with increasing altitude, such as some Coleoptera (Smith *et al.,* 2000; Chown & Klok, 2003), Hymenoptera (Ruttner *et al.,* 2000), and Orthoptera (Rourke, 2000), among others. Opposite patterns, of declining body size with increasing elevational have also been found in Coleoptera (Chown & Klok, 2003) and Orthoptera species (Bidau & Marti, 2007), to mention only a few. Moreover, other insect species show no significant variation in body size along altitudinal gradients, such as some species of geometrid moths (Brehm & Fiedler, 2004) or Coleoptera (Chown & Klok, 2003).

Overall, body size is considered a complex life-history trait and its variation seems to be determined by a range of factors, including the temperature (Kingsolver & Huey, 2008), growth rate and the mass gained during the developmental period (Davidowitz *et al.,* 2004). However, temperature (Angilletta *et al.,* 2004) and seasonality of resource availability (Chown & Klok, 2003) seem to be the main drivers of this variation. Nevertheless, body size influences the spatial scale at which organisms manage to access resources (Petchey & Gaston, 2006), affecting the species turnover at the habitat scale, but also along elevation gradients and among regions (Mason *et al.,* 2007). Species with life-history traits of this level of flexibility are potentially sensitive to environmental disturbances, most probably leading to shifts in community composition or even species losses as a result of environmental change, affecting the entire ecosystem functionality (Petchey & Gaston, 2006).

Wild bees (Hymenoptera: Anthophila) were chosen as a model group for this study given their ecological and economical importance as pollinators and their sensitivity to anthropogenic disturbance (Biesmeijer *et al.,* 2006). Human induced disturbances known to affect bee abundance include population isolation through habitat fragmentation (Winfree *et al.,* 2009), agricultural intensity (Winfree *et al.,* 2009), fire (Potts *et al.,* 2006), grazing (Vazquez & Simberloff, 2002), climate change (Kerr *et al.,* 2015), and exposure to chemicals such as pesticides (Shuler *et al.,* 2005). Thus, the identification of life-history traits that affect species sensitivity to environmental disturbances will potentially support the prediction of community responses to such changes.

The diversity of studies on life-history traits in bees, particularly body-size, have essentially investigated the correlation between size and: i) the species' robustness regarding loading capacity (of pollen and nectar) and reproductive rate and success (Giovanetti & Lasso, 2005; Bosch & Vicens, 2005) ii) thermoregulatory ability (Stone, 1993; Pereboom & Biesmeijer, 2003) iii) flight distance capacity and consequently foraging range (Araújo *et al.*, 2004; Greenleaf *et al.*, 2007; Torné-Noguera, 2014) and iv) the influence of size on bee community assemblages in fragmented habitats (Wray *et al.*, 2013).

Limited research has been applied to the effects of body size and other ecological traits of bee responses to anthropogenic disturbance, but the limited literature reveals that body size inconsistently affects the way species react to environmental changes (Williams *et al.*, 2010). As far as could be determined, there are only a few studies that investigated the variation in body size of bee species along altitudinal gradients.

In the temperate climate of the Elbrus mountains in Russia, along a gradient from 50 to 2200 m.a.s.l., the body size of an important pollinator (*Apis mellifera meda*) was investigated and the authors concluded that it followed Bergmann's rule of increasing body size with increasing elevation (Rutner *et al.*, 2000). A study conducted in the Norwegian mountains between 240 and 1100 m.a.s.l. indicated that bee body size was larger at high elevation sites when compared to populations found in low elevation sites (Maad *et al.*, 2013). In Germany, a study on the variation on bee's morphometric traits along an altitudinal gradient between 641 and 2032 m.a.s.l. revealed that body size increased with increasing elevation, both at the community and species level (Peters *et al.*, 2016). A study with the three species of *Bombus* in Great Britain also revealed a variation in body size consistent with Bergmann's rule (Scriven *et al.*, 2016).

In Ethiopia, the variation of honey bees body size along an elevational gradient was investigated and an increase in body size with altitude was found (Meixner *et al.*, 2010). Furthermore, studies performed in Mount Kilimanjaro revealed an overall decrease in bee species of large body size with increasing altitude (Schellenberger Costa *et al.*, 2017; Classen *et al.*, 2017), although individuals within species increased were found to be on average larger with increasing elevation (Classen *et al.*, 2017).

Overall, the response of bee body size to increasing altitude in temperate zones matches Bergman's rule. In these regions, there is a sharp decrease in temperature with altitude, and an increase in body size would be expected to have thermoregulatory benefits as the ability to elevate the thoracic temperature above the air temperature increases with body size (Stone & Wilmer, 1989), enhancing the flight and foraging performance in cooler climates and favouring the colonization of higher elevations. In tropical zones similar temperature drops occurs with increasing altitude (Schellenberger-Costa *et al.,* 2017; Classen *et al.*, 2017) but the individuals body size variation has been reported to increase with increasing altitude while the overall body size within communities has been reported to decrease with increasing elevation (Classen *et al.*, 2017). The steep drop in temperature with elevation is likely more pronounced across gradients with higher maximum altitudes like the Kilimanjaro gradient (summit at 4,000 m.a.s.l.), i.e. which are greater than the 1651 m summit in this study, although the temperature drop per elevation is similar (~ 6.1C° per 1000m in Mount Kilimanjaro compared to a decline in ~ 6.0C° per 900 m in the Bruco gradient).

Environmental variables associated with altitude, such as precipitation, have also been found to indirectly impact bee body size due to the temporary shortage or abundance of foraging resources (Roubik, 2001; Peruquetti, 2009). In particular, higher rainfall levels have been associated with increased food availability and increased bee body size (Peruquetti, 2009).

In this study, the relationship of wild bee body size with altitude (altitudinal range from 760 m a.s.l. to 1651 m a.s.l.) and associated variables such as rainfall along a tropical altitudinal gradient was investigated. I expect that the analysis of how altitude and rainfall gradients associated with the elevational transect affect bee body size will reveal how the variation in the pattern of this life-history trait is expressed at the community and intraspecific levels. If Bergmann's rule can be demonstrated along these gradients, then this would be the first demonstration of its applicability at different levels of biotic organization in tropical bee communities.

Following the results of other studies of insect body size variation along tropical altitudinal gradients, I hypothesize that:

1. Body size will increase with altitude (assuming there are physiological constraints imposed by lower temperatures of the higher elevations of the upper stations of the Bruco gradient; Schellenberger Costa *et al.*, 2017). Temperature gradients may only be operational in winter, or at dusk and dawn – periods where ambient temperatures are likely to be at their lowest.

4.2. METHODOLOGY

For a description of the study area, sites and sampling design refer to section 3.2. (pp. 55).

4.2.1. BODY SIZE MEASUREMENTS

One measure of body size – body length – was investigated along the altitudinal gradient. Body length (to the nearest 0.01 mm) was measured from a midpoint on the frons to the last visible abdominal segment, using the manual line tool (Leica EZ4 HD software), from digital images captured with a calibrated Leica EZ4 HD photomicroscope, following Malo & Baonza (2002). Where a specimen was either larger than the field of view at lowest magnification, or had a reflexed abdomen, multiple images were taken, and the individual measurements then summed to give the entire body length.

4.2.2. STATION ALTITUDE AND RAINFALL DETERMINATION

The 12 sampling sites were situated from the lowest (760m.a.s.l.) to the highest (1651m.a.s.l.) altitudes of the transect, positioned according to groups of four points – four at the lower altitude, four in the middle and four at the higher elevations. Altitude was recorded in the field using a handheld gps (Garmin model Oregon 450) with barometrical altimeter accuracy of ±3m.

Measures of annual precipitation (mm) and mean annual temperature (°C), were obtained for each site using Worldclim version 1.4 (1960-1990). I then tested for effects of these environmental variables on bee body size along the altitudinal transect. The results from correlating these environmental variables and altitude (Appendix III) indicated that annual mean temperature declined significantly with increasing elevation (r= - 0.95, p< 0.05) and annual precipitation increased significantly with increasing elevation (r= 0.93, p< 0,05).

4.2.3. STATISTICAL ANALYSES

Linear regression was used to explore the relationship between mean body size of all bee genera (per the 12 sites) and altitude. Given the small number of site means, the relationship between altitude and mean bee body size was examined by partitioning the data into a low (sites 1-6) and high (7-12) altitude group, for both community (all genera) and intraspecific level (for the most abundant species that was represented in reasonable numbers at all sites *Macrogalea candida*). The means from the two groups were compared using an independent sample t-test. Normality of the two datasets was confirmed using a Shapiro-Wilk test.
A further analysis of variance (one-way ANOVA) was conducted to compare variation among low (sites 1-4), middle (sites 5-8) and high (sites 9-12) elevation groups. Tukey post hoc tests were conducted to determine which groups had significantly different mean body lengths.

All statistical procedures were performed using the computer software STATISTICA version 13.5 and the results are presented in agreement with APA guidelines (APA, 2010).

4.3. RESULTS & DISCUSSION

In total, 843 bee individuals were caught with the different sampling methods and between 10 and 20 genera per site. As can be observed in Figure 4.1, outliers were found in most sites. This was not unexpected, given the considerable differences in size between large-bodied genera such as *Xylocopa* and the very small meliponine bees, specifically *Hypotrigona*.

At the community level (all individuals of all genera), body size increased significantly with increasing elevation (Figure 4.2) (r = 0.79, p < 0.05, N = 12). The increase in mean body size was partly a consequence of greater abundances of larger-bodied genera at the upper stations of the altitudinal gradient, e.g. *Megachile and Xylocopa*.

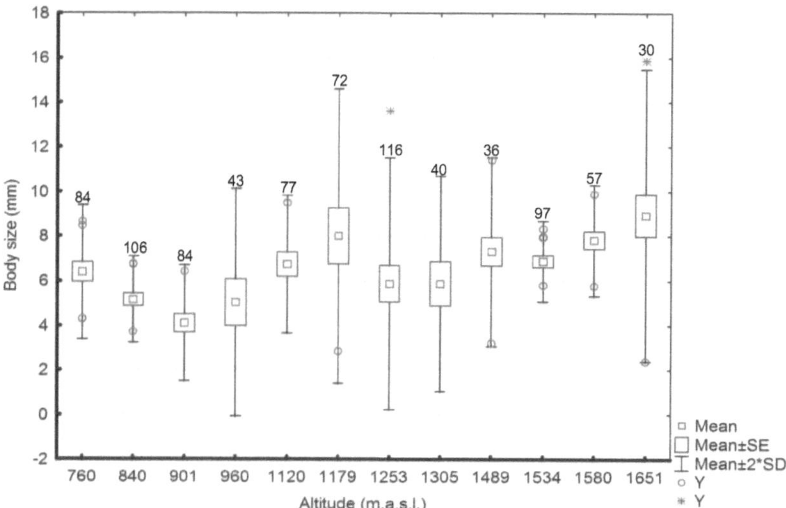

Figure 4.1 – Bee genus body size variation along an altitudinal gradient. Points above and below rectangles (mean SE) indicate outliers and extreme outliers. Numbers of total specimens collected per site are presented on top of whiskers.

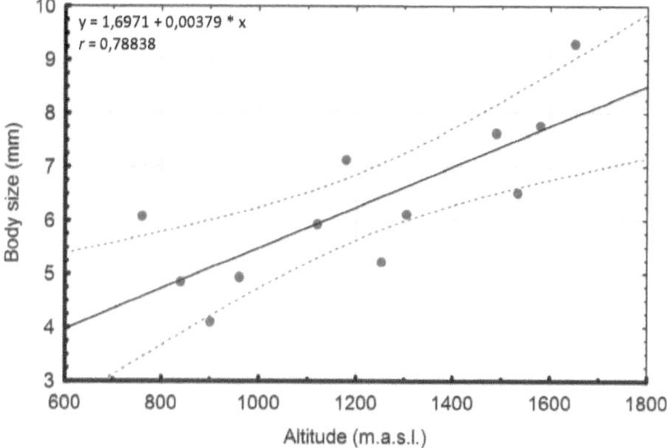

Figure 4.2 – Regression of mean bee body size with elevation at the community level. Along the gradient, bee body size increased with increasing altitude.

Temperature dropped by 6 degrees at the upper stations when compared to the semi-arid, lower part of the transect., with annual mean temperatures of 24°C at the lower altitude of the transect against 18°C at the higher altitude. The correlation of larger body size with altitude is in line with

Bergmann's rule (viz. mean body size being larger in colder habitats) (Figure 4.2), but this relationship does not necessarily imply causality, as other environmental factors may act synergistically with temperature in driving these changes in body size.

Bee body size was positively correlated with annual precipitation, another environmental variable known to affect bee body size (Peruquetti, 2003). Along the transect of the study area, precipitation increased with increasing altitude (528mm in the lowest site to 717mm in the highest site) along with bee mean body size (Figure 4.3). Higher rainfall levels are associated with increased food availability which is thought to play a significant and positive role in bee body size (Periquetti, 2003). In line with the relationship between increased body size with altitude, body size had a significant, negative relationship with temperature (Figure 4.3).

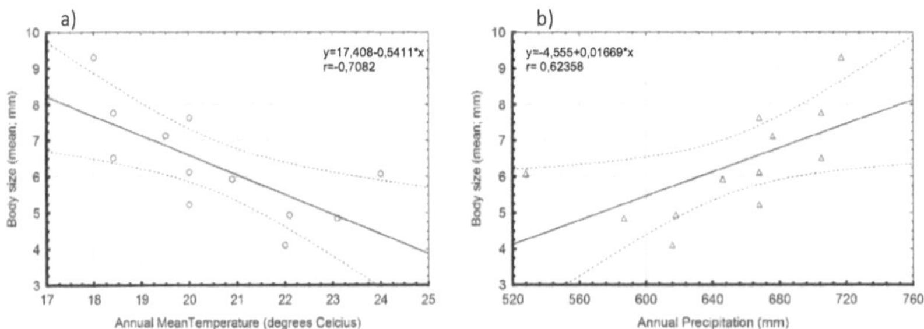

Figure 4.3 - Change in bee body size (mean) with a) temperature and b) precipitation along the altitudinal gradient. Bee body size increased along the transect presenting higher values in cooler habitats and increased with increasing rainfall.

The independent samples t-test used to compare whether mean bee body size in higher elevations (1253-1651) was larger than for low elevations (760-1179) showed that the mean body size of the lower elevation group (*mean* = 5.807; *SD* = 2.139) differed significantly from that of the higher elevation group (*mean* = 7.205; *SD* = 2.462); $t(116)$ = -3.269, p = 0,001). These results support the positive correlation of body size of the bee communities with altitude, as determined by the regressions.

An analysis of variance (one-way ANOVA) conducted to compare variation among low (sites 1-4), middle (sites 5-8) and high (sites 9-12) elevation groups showed that there was a significant difference in mean body size between groups [$F(2, 840)$ = 81.626, p = 0.001] (Figure 4.4).

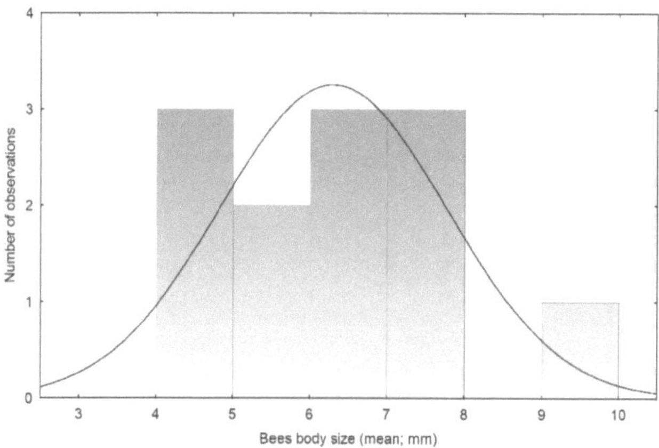

Figure 4.4 - Shapiro-Wilk normality test result [W = 0.97; p = 0.90]. Histogram for the distribution of all bee genera mean body size (mm).

The post-hoc Tukey test showed that bees from high elevation sites (1489 – 1651m.a.s.l.) presented a larger body size (*mean* = 7.90mm, *SE* = 0.172, *N* = 221) differing significantly ($p <$ 0.001) from bees in low elevation sites (760 – 960m.a.s.l.; *mean* = 5.07mm, *SE* = 0.144, *N* = 317) and from bees in middle elevation sites (1120 – 1305m.a.s.l.; *mean* = 6.58mm, *SE* = 0.146, *N* = 305).

In order to analyze variation in body size at the intraspecific level, *Macrogalea candida* was chosen as it proved to be present at all sampling sites and in sufficient abundances to allow statistical analyses (Figure 4.5). At the species level, body size also increased significantly with increasing elevation (Figure 6) (r = 0.578 for $p <$ 0.05, *N* = 12).

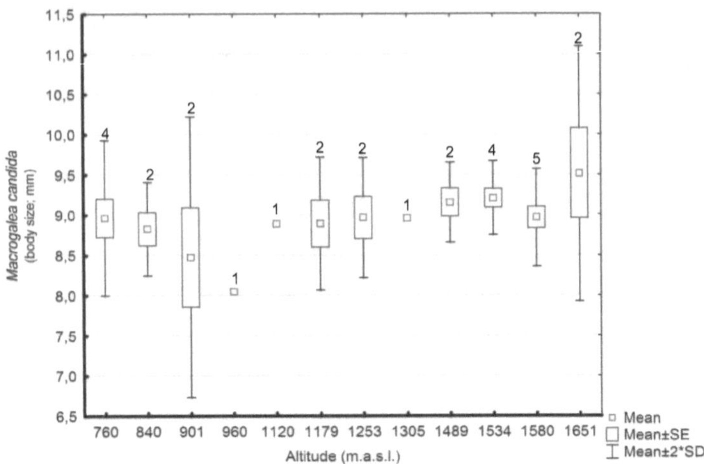

Figure 4.5 – Macrogalea candida body size variation along an altitudinal gradient. Numbers of total specimens collected per site are presented on top of whiskers.

An independent samples t-test was conducted to compare whether at the intraspecific level, mean bee body size of higher elevation populations (1253-1651 m.a.s.l.) was larger than of low elevation populations (760-1179 m.a.s.l.). There was a significant difference between the mean size of the lower elevation group (*mean* = 8.763; *SD* = 0.350) and the higher elevation group (*mean* = 9.119; *SD* = 0.132); $t(10)$ = -2.329, p = 0.04).

A further analysis of variance (one-way ANOVA) was conducted to compare variation of body size of *M. candida* among low, middle and high-altitude groups (as was done for all genera). Assumptions at normality checks (Shapiro-Wilk test, Figure 4.4) were met. There was no significant difference in mean body size between groups [$F(2, 90)$ = 1.46, p = 0.239] (Figure 4.6).

Positive associations of increasing body size with altitude were evident here in spite of the relatively low number of collected specimens. The distribution of wild bee body size along an altitudinal (and rainfall) gradient in Bruco – Serra da Chela, Angola – gives evidence of an increase in size with increasing elevation at different levels of biotic organization. These patterns are in agreement with Bergmann's rule that explains patterns of variation in an organism's body size by physiological constraints. The findings of this study indicate that, on average, all bee genera (=

communities) and one species present larger body size in cooler habitats with higher annual precipitation.

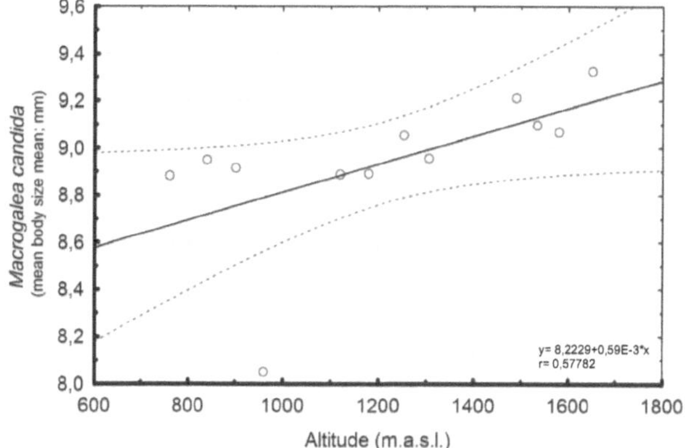

Figure 4.6 - Change in mean body size of Macrogalea candida with elevation at the species level. Along the gradient, the body size increased steadily with increasing altitude.

The limited research on bees' body size variation with altitude, both in temperate or tropical regions, indicates an association between larger sized bees with high elevation, low temperature and higher rainfall environments as this trait favours bee foraging activity over a broader thermal range (Peters *et al.*, 2016; Classen *et al.*, 2017). The results from this study also substantiate the existence of a pattern of increasing body size with altitude, in a tropical zone and along a transect that experiences a steep drop of the annual mean temperature between the lower altitude (24°C) and summit (18°C). The existence of larger organisms that present low surface-to-volume ratios in cooler habitats is known to be energetically favoured due to the reduced heat loss through conduction (Meiri & Dayan, 2003), and it could be an explanation for the pattern observed in this study as the decline in temperature between the extremes of the transect is steep. This effect is expected to be even more pronounced during the winter (dry season) when mean temperatures at the lower altitude of the transect are lower than in summer (21°C) as are the ones in the summit (15.4°C).

Other factors, that were not assessed in this study, could contribute for the increase of bee body size along the altitudinal gradient, namely mechanisms of adaptive plasticity through prolonged development times or delayed maturity (Anguilletta *et al.*, 2004). The thermal performance of bees is also relevantly influenced by other morphological traits that were not evaluated in this analysis, such as density of thoracic hairs, length of appendages, melanisation or body hairs (Hodkinson, 2005), and it is possible that these anatomical responses to temperature were present in the bees along the gradient. *M. candida* is an ideal species where this can be studied, as it demonstrated a body increase with altitude, and was present at all sites along the altitudinal transect.

Resource availability has also been presented as an influencing factor for bee body size variation along altitudinal gradients and was not considered in this study. Flower resource reduction along increasing elevations (Körner, 2000) restricts the energy supply available at high altitudes, consequently lowering harvesting rates in cooler habitats (Classen *et al.*, 2015) and sustaining only smaller populations of larger organisms (Damuth, 1987).

However, a potential caveat to these results is the method applied to measure bee body size. According to Cane (1987), "body length has allometric drawbacks, particularly for the more elongate and cylindrical bodies of stem-nesting taxa", especially considering that the length of the metasoma is not definable because terga and sterna can easily slide in and out, potentially introducing errors in the measurements. For larger bees, the variation in length can be a millimetre or more, which is a hundred times higher than the precision of the measurement (i.e. 0.01mm). Measuring bee body size using intertegular distance could reduce the errors associated with measurements as this is considered a robust estimator of body mass (Cane, 1987).

Morphological trait variations patterns along altitudinal gradients are of ecological relevance as they reflect the organism's or community's capacity to overcome changes in the environmental conditions. The apparent ability of *M. candida* to change body size in response to environmental conditions suggest some degree of developmental plasticity, at least in response to cold.

Overall, body size of bee communities along an altitudinal gradient in Bruco showed an increase with increasing elevation. To my best knowledge, this constitutes the first demonstration of single species and community level body size increase with increasing elevation in beesin the tropics. Further work, considering other morphological traits as well as environment factors (e.g. resource availability) may improve the understanding about organism and community responses to a changing climate.

CHAPTER 5. CONCLUSIONS

5.1. GENERAL STATUS OF BIODIVERSITY IN ANGOLA

The Angolan civil war lasted for twenty-seven years (1975-2002) with the associated military interference compromising infrastructure and scientific research. Additionally, the post-war period was marked by severe financial constraints that contributed to a 40 year gap in knowledge in all biological fields. Recently, the Angolan Ministry of Environment, in partnership with international organizations and some other initiatives, has been making efforts to gather contemporary data on the status of Angola's biodiversity.

As expected, vertebrate research and inventorying has taken priority over that of invertebrates. Profiles for reptiles and amphibians are receiving adequate attention and preliminary checklists for national parks (Ceríaco *et al.*, 2016) and comprehensive atlases of Angola (Marques *et al.*, 2018) have been created. Special attention has been given to birds mainly in areas of high endemism such as Kumbira forest and Mount Moco, even in so far as the publication of field-guides (Mills, 2018) and a preliminary checklist for the country, comprising 940 species (Mills & Melo, 2013). Preliminary checklists were also generated for two national parks (Mills *et al.*, 2008; Hines, 2018a). As for terrestrial large and medium sized mammals, preliminary checklists have been created for six of the eight existing national parks and the reserve protecting the natural habitat of the giant sable (Funston *et al.*, 2017; Overton *et al.*, 2017; Groom *et al.*, 2018; Elizalde *et al.*, 2019). Moreover, inventories are available for the country based on online information and recent surveys (Taylor *et al.*, 2018) and a biogeographical regionalization has been done (Rodrigues *et al.*, 2015). Surprisingly, even a new species of primate has been discovered (Svensson *et al.*, 2017), outlining the need for more detailed studies in this biodiversity-rich country. National vegetation checklists were created based on information collected in natural history collections (Figueiredo, 2009) and updated with recent field work (Goyder *et al.*, 2019). Preliminary checklists for the various national parks (Hines, 2018b) are also informing conservation policy. A biogeographic survey of freshwater fishes has recently been created (Skelton, 2019) based on a preliminary checklist (Stiassny *et al.*, 2007) and recent field work.

In contrast, insects have generally been neglected until recently, when information was gathered from natural history collections on butterflies and moths, allowing for the compilation of a national checklist with 792 species and subspecies (Mendes *et al.*, 2019). The first insect survey ever conducted in an Angolan national park (Bicuar) with emphasis on Lepidoptera (Picker, 2018) produced a preliminary checklist of 94 species. Data on carabid beetles from recent surveys resulted in a preliminary checklist and the description of new species (Serrano *et al.*, 2017). Odonata is one insect order that has been more comprehensively studied in Angola , with recent focus on biogeography and endemism of the fauna, generating a National checklist of 264 species (Kipping *et al.*, 2017; Kipping *et al.*, 2019), but also including the discovery of at least 27 new records for Angola in 2018 alone, several of which are undescribed (Kipping *pers.com.),* suggesting that Angola might support the richest Odonata fauna in Africa.

Angola is incredibly rich in terms of biomes and ecoregions, as well as climatic and biological diversity. Sadly, most of the contemporary research has been opportunistic, following the same pattern of the colonial period. Data available on online platforms, mainly from natural history collections, indicate that nearly half of the northern part of country (northwest) has been poorly surveyed – the same deficiency holds for almost two-thirds of the inaccessible eastern parts. Recent vertebrate surveys have also focused their efforts on the southwest (Overton *et al.*, 2017; Marques *et al.*, 2018) and southeast of the country (Funston *et al.*, 2017) or along the escarpment (Mills, 2018). An ongoing biodiversity survey, starting at the central plateau and following towards the southeast of the country led by the Okavango Wilderness Project, is generating contemporary data on a region of the country that was barely known. The constant discovery of new species to science or new records for the country (Serrano *et al.*, 2017; Svensson *et al.*, 2017; Marques *et al.*, 2018; Kipping *et al.*, 2019; Skelton, 2019) outline the understudied diversity of Angola.

5.2.   PROFILE OF HYMENOPTERA: ANTHOPHILA IN ANGOLA

Knowledge on the diversity of Hymenoptera - Anthophila of Angola presents the same pattern as for other biological fields: outdated, reliance on opportunistic surveys and limited geographical coverage where only a small fraction of the country's area has been surveyed – not surprisingly, many entire biomes and ecoregions are still to be surveyed.

By digitizing information from only one of the two entomological collections in Angola, the present study added 1436 records of bee genera to the 928 existing in the global platform GBIF (https://www.gbif.org), increasing the records almost 155%. These data augmented the number of historical records for 14 of the 18 provinces of the country and added records from two new provinces. Still, two provinces (Zaire and Cabinda) completely lack historical records and nearly half of the country has been very poorly surveyed. This highlights the need for taxonomic specialists to work on historical collections, either the existing in the country or those hosted in foreign countries, in order to gather all possible historical information on species distribution. However, once data on specimens housed at insect collections in European museums is digitized, especially the Portuguese that hold a great number of specimens from the pre-independence expeditions, this numbers will probably increase but the observed collector bias or geographical coverage are expected to be the same, as it has been observed for other biological groups (e.g. Marques *et al.*, 2018).

The historical bibliography from the pre-independence period that is not available in the country should also be accessed. Old literature housed in Angolan libraries is scarce, uncatalogued and access is extremely bureaucratic, as application for permits to access collections must be made at least a month in advance, as well as restricted because the access is limited to the library room and no copies are allowed, so information must be retrieved instantly). In spite of these restrictions on accessing historical data, the bibliographic revision in Chapter 2 resulted in the most up to date checklist of the bees of Angola, comprising 186 species. During the timeframe of this study, the preliminary checklist has already been updated with the obtained results adding 23 new species records for the country and 6 genera. Therefore, the preliminary checklist of bees of Angola is now comprised of 209 species from 47 genera and five families.

Nonetheless, the present study was limited by the lack of access to one important Angolan entomological collection – the Dundo Museum - and several entomological collections with type material from Angola held overseas, as well as by the unavailability of some of the historic literature. This information should be made publicly available and incorporated into any future profiles of Angolan bees.

A more structured sampling for Angolan bee diversity needs to be implemented. Apart from distributional data and improved taxonomic surveys, information should also be gathered on the pollination biology of the more important taxa, given their importance for rural crop pollination (Rodger *et al.*, 2004). For a national inventory of the bees, the example of other successful regional surveys (e.g. Coaton & Seasby, 1972) could be followed. The survey could realistically sample half degree cells where one of four adjacent quarter degree cells in each half degree cells being sampled. This would generate presence data (there would be ~ 487 quarter degree cells needing to be surveyed, pending on accessibility and other limitations such as the presence of land-mines). Standardized collecting methods with equal sampling intensity/site (sweep transects, pan and malaise traps) would need to be employed. Pending on resources, sampling could be undertaken in both dry and rainy seasons. Nevertheless, high frequency sampling (every 7-10 days), particularly in spring when bee diversity and density reach a peak, is important for the collection of rare species, or species that are only active over short periods (Banaszak *et al.*, 2014). Protocols of cooperation should be implemented with taxonomic experts around the world so that collected specimens could be outsourced for identification. In order to support future taxonomy and systematical work in Angola, type material could be hosted at the Natural History Museum in Luanda to serve as reference for species identification.

In conservation planning, individual taxonomic groups are frequently used as surrogates for total biodiversity, although for a reliable proxy of total species richness a range of biome-specific taxa should be used (Larsen *et al.,* 2009). Knowledge on Angolan Odonata diversity and distribution (Kipping *et al.,* 2017) is advanced but this group is likely to serve mostly as a surrogate taxon for other freshwater taxa. Bees being a monophyletic group of ecologically important insects, with their high richness and abundance (Klein *et al.,* 2003) are well-suited as terrestrial habitat bioindicators (Tscharntke *et al.,* 1998; Tylianakis *et al.,* 2004; Sheffield *et al.,* 2013). The group is sensitive to environmental disturbances (Biesmeijer *et al.,* 2006), relies upon local plant communities and comprises an assemblage of species with varying social structure, nesting guilds are life history traits (Williams *et al.,* 2010). Additionaly, bees are likely more susceptible to negative effects of small effective population size (Zayed & Packer, 2005). These determine the individual and community response to environmental changes. Given the limited resources and the comparatively well-studied status of Angolan bees (certainly in comparison to other insect taxa) they would appear to be a suitable surrogate group for other insects and a national inventory and spatially explicit database of the taxa would assist in the identification of areas with exceptional richness or endemism.

5.3. ALTITUDINAL GRADIENTS AND BEE COMMUNITIES

In nature, ecosystem services such as pollination are provided by a number of species that increases monotonically with increasing spatial scale, as different species perform the same functions in different locations (Anderson *et al.*, 2011). Species turnover across habitats is particularly evident in heterogeneous landscapes such as those existing along large altitudinal gradients (Paritsis & Aizen, 2008), because bee species richness and assemblage composition are affected by species interactions, resources and nesting site availability (Tscharntke *et al.*, 2005). Another factor that contributes to the ecosystem service provision is the fact that natural ecological communities are composed by a few dominant species, which are in numerical advantage and provide most of the functionality (Kleijn *et al.*, 2015), with the numerous rare species (McGill *et al.*, 2007) contributing significantly to pollination at specific locations and possibly being oligolectic and essential to a single plant taxon (Kremen *et al.*, 2002). The effects of species turnover and dominance are thus felt in parallel (Anderson *et al.*, 2011).

Altitudinal gradients have been used as model systems for changes in the climatic conditions in an attempt to analyze the effect of environmental variables in bee community assemblages, both in temperate (Hoiss *et al.*, 2012) and tropical (Perillo *et al.*, 2017) regions across small spatial scales. A pattern of linear decline in species richness and abundance with increasing altitude was found in both regions (Hoiss *et al.*, 2012; Perillo *et al.*, 2017). Along tropical altitudinal gradients, variation in environmental variables such as altitude, precipitation, topography and soil type regulate the distribution of vegetation (Holland & Steyn, 1975), which in turn is expected to affect bee community assemblage. Temperate gradients are more likely to impose direct physiological constraints on species diversity and community composition. While bee species richness and life-history can be explained by a combination of biotic and abiotic factors, temperature seems to be a main contributor both in temperate and tropical environments (Hoiss *et al.*, 2012; Classen *et al.*, 2015). Bees are expected to suffer strong selection pressure at high altitudes due to the cooler conditions, experiencing physiological changes in life-history traits such as body size or coloration in order to improve thermoregulation (Hoiss *et al.,* 2012; Peters *et al.,* 2016).

In contrast to the results obtained in other studies in the tropics, I found an increase in bee diversity (carried out at the generic level) with increasing altitude. In addition, community structure differed along the gradient, with distinctive communities in low elevation sites being dominated by *Braunsapis, Ceratina, Hypotrigona* and *Anthidiellum,* and high elevation sites the distinctive communities were dominated by *Amegilla, Macrogalea, Patellapis* and *Xylocopa.* This result might be consistent with findings from studies of other taxa that attribute to the Angolan escarpment an importance as center of endemism and speciation (e.g. Kipping *et al.,* 2017; Dean *et al.,* 2019). Angola's topography is generally characterized by i) the coastal lowland strip that lies below 200m; ii) the western mountain chain that abruptly rises 1000m; iii) the central plateau, between 1400 and 1800m; and iv) a few peaks like Mount Moco (2620m), Serra da Neve (2400), or Mount Namba (2200) (Diniz, 2006). The escarpment mountain chain is thought to create a barrier between the species adapted to the arid environments of the coastal lowlands and the ones adapted to the miombo woodlands that cover the central plateau, therefore creating a steep ecological gradient with high species turnover (Hall, 1960).

The results from this study are possibly explained by a combination of factors (Classen *et al.,* 2015) where the exceptional geomorphology and landscape shape the (direct) influence of environmental variables such as temperature and precipitation on the vegetation communities along the gradient therefore (indirectly) affecting bee community assemblages, particularly for the more specialized bee genera and species (Wcislo, 1996).

5.4.   BEES BODY SIZE VARIATION ALONG ALTITUDINAL GRADIENTS

Globally, seasonal variation in ambient temperature is greater in temperate regions when compared to the tropics, and a similar pattern is observed for temperature overlap between different altitudes along elevation gradients, with greater overlap between sites in temperate locations (Ghalambor *et al.,* 2006). Additionally, daily oscillations in temperature are also larger in high-altitude locations in the tropical zones than the fluctuations observed in temperate climates, for the same altitude (Ghalambor *et al.,* 2006). Therefore, the contrast in temperature between sites at the extremes of tropical elevation gradients acts as a physiological barrier, favoring the presence of organisms with higher tolerance to temperature variation in cooler climatic conditions and, consequently, increasing species turnover along the gradient (Ghalambor *et al.*, 2006). However, in addition to turnover, some species have sufficient phenotypic plasticity in life-history traits - such as body size - to adapt to thermal constraints along the gradient, including those imposing access to resources (e.g. foraging) (Mason *et al.*, 2007).

Smaller organisms experience faster rates of heat loss than do large bodied organisms, and are thus less tolerant of cooler habitats (Peters *et al.*, 2016) as the energy costs associated with thermoregulation will be higher (Stabentheiner & Kovac, 2014) and foraging efficacy reduced and limited to periods of higher temperatures (Buckley & Kingsolver, 2012). When weather conditions are optimal, bees tend to forage in sunshine using solar radiation to warm up (Kovac & Stabentheiner, 2011).

Most bee species overcome these constraints of ectothermic organisms with varying degrees of endothermy - some producing heat with the thoracic muscles, elevating thoracic temperature above the ambient temperature, allowing them to explore the resources even at cooler temperatures or overcast conditions (Stabentheiner *et al.*, 2012). Small bodied bees are thus still at a disadvantage in colder environments as they suffer tremendous heat loss when foraging at low temperatures, while large bodied bees are not just faster at increasing their thoracic temperatures but also have a slower heat loss (Bishop & Armbruster, 1999). Thermoregulatory capacity of bee species belonging to different families has been proved to be related to body size since large bodied bees have higher thermoregulatory ability and start foraging earlier and at lower temperatures than small bodied species (Bishop & Armbruster, 1999), increasing the resource exploitation as nectar levels are usually higher in the early morning (Wright, 1988). Life-history traits such as body size and thermal physiology have been shown to significantly contribute to the performance of bees in cold climatic conditions and consequently strongly influence the phenology in community structure (Osorio-Canadas *et al.*, 2016). Thus, bee communities are expected to be composed of organisms with different thermoregulatory abilities comprising large bodied endothermic species and small bodied ectothermic solitary species (Bishop & Armbruster, 1999). The large bodied species (above a determined threshold) are expected to follow

Bergmann's rule, with small bodied bees following different patterns (Osorio-Canadas *et al.*, 2016).

Limited research has been done on the variation of bee body size along altitudinal gradients, but the few studies that have been carried out indicate that future changes (warming) in climatic conditions will most probably alter bees body size (by reduction), consequently affecting their thermoregulatory capacity and, at a larger scale, the overall patterns of community assembly (Osorio-Canadas *et al.*, 2016). Even along the tropical altitudinal gradient of this study, bee body size was found to increase with altitude by 58% (mean length at lower altitude = 5.06mm against mean = 7.99mm at higher altitude elevation sites) at the community level but also at the intra-specific level, with *Macrogalea candida* increasing 4% in size between the lower altitude and summit of the transect (mean length at lower altitude = 8.76mm against mean = 9.12mm at higher altitude elevation sites), suggesting that the fairly small temperature gradient along the transect was sufficient to select for body size – either directly, but more likely through multiple avenues, including indirect effects such as changes in rainfall, plant community composition and plant phenology.

Studies of variation of body size at the intra-specific level and along altitudinal gradients, where the number of influencing variables is more limited when compared to coarser geographical scales, could be useful to contribute to the understanding of the underlying factors that affect turnover and change in community structure. Future research on the variation of bee body size should then focus on a single and common species, evaluate its thermal physiology and investigate the existence of patterns of physiological adaptions along a gradient of altitude.

5.5.   GENERAL CONCLUSIONS

Wild bees are important pollinators for both the native plant community and mass flowering agricultural crops (Carvalheiro *et al.*, 2011). Yet, the factors that generate the diversity and composition of their communities are poorly-known for tropical systems. E.g. community structure is strongly influenced by the landscape context in the tropical regions, increasing the abundance of eusocial bee species at the forest edge in contrast to a decrease on deforested areas

(Brosi *et al.*, 2007). Conservation of native forest is critical for the sustainability of wild bee diversity, assuring an efficient pollination of native plants while contributing to an improved pollination of agricultural crops (Carvalheiro *et al.*, 2011, 2012).

The Afromontane forest existing along the altitudinal transect of Bruco pass has been mostly well preserved along the years since the area is extremely inaccessible. The recent (2012) rehabilitation of an old trail that connects the higher altitude and lower altitude of the transect, as well as the creation of a technical agriculture school at the lower altitude and rehabilitation of one in the higher altitude, have provided better access to the area and increased the livelihood alternatives for the few small villagers who live along the transect (e.g. charcoal exploration, cattle grazing, etc.). The effects of these activities are already evident as both schools have a high demand for agricultural products and charcoal which is severely contributing for a local increase in deforestation. A land management program that combines the natural forest with flowering crops could improve productivity while reducing the need to expand the area occupied by crops (Carvalheiro *et al.*, 2011) - while at the same time assuring livelihood sustainability to those isolated populations. This could be accompanied by the implementation of efficient cookstoves in both schools and within the population, in order to reduce charcoal dependency (Johnson & Chiang, 2015).

Effective conservation planning is highly dependent on robust, spatially explicit biodiversity data. The generation of comprehensive insect profiles through long term monitoring studies at the national level can identify areas with high species turnover and endemism – patterns that could be missed if only vertebrate data were considered. In under-developed countries like Angola, where a poorly documented but highly rich biodiversity is severely threatened by unregulated land use, and research is hampered by financial constraints and lack of human resources, the combined use of indicator taxa for both terrestrial (bees) and freshwater (Odonata) environments might be a reasonable surrogate for total biodiversity status assessments.

The implementation of cooperation programs with experts from regional (e.g. the Plant protection research institute - Agricultural research council, South Africa; Iziko Museum – Cape Town) or international institutions (e.g. Smithsonian Institution - USA) could improve national capacity, as would the integration of initiatives such as the African Pollinator Initiative (API). A successful example of how such cooperation contributes to enhance the scientific knowledge on Angola's biodiversity and allows for informed decision-making processes is the Okavango Wilderness Project. Through several expeditions starting in 2015, an interdisciplinary team of scientists and explorers has gathered data on the biodiversity and socio-economic profile of resident the human population of the Okavango watershed, not only making exciting scientific discoveries (e.g. potentially 34 new species to science, actualization of distribution ranges, among others; Okavango wilderness project, 2017) but also inducing the creation of a new conservation area in Angola with ~ 72,000km^2.

Angola's is unique in ecosystem diversity and for keeping vast wilderness areas, with pristine habitats and an incredibly rich biodiversity. Further research in this challenging but exciting country is urgently needed in all biological fields to contribute for effective and scientifically informed conservation planning.

REFERENCES

Abrahamczyk S, Kluge J, Gareca Y, Reichle S, Kessler M (2011) The influence of climatic seasonality on the diversity of different tropical pollinator groups. *PLoS ONE* 6(11): e27115.

Allsopp MH, de Lange WJ, Veldtman R (2008) Valuing insect pollination services with cost of replacement. *PLoS ONE* 3(9): e3128.

Almeida EAB, Pie MR, Brady SG, Danforth BN (2012) Biogeography and diversification of colletid bees (Hymenoptera: Colletidae): Emerging patterns from the southern end of the world. *Journal of Biogeography* 39(3): 526–544.

Almeida EAB, Packer L, Melo GAR, Danforth BN, Cardinal SC, Quinteiro FB, Pie MR (2018) The diversification of neopasiphaeine bees during the Cenozoic (Hymenoptera: Colletidae). Zoologica Scripta 2018: 1-17.

Anderson MJ, Crist TO, Chase JM, Vellend M, Inouye BD, Freestone Al *et al.* (2011) Navigating the multiple meanings of β diversity: a roadmap for the practicing ecologist. *Ecology Letters* 14: 19-28.

Angilletta MJ, Steury T, Sears M (2004) Temperature, growth rate, and body size in ectotherms: fitting pieces of a life-history puzzle. *Integrative and Comparative Biology* 44:498–509.

APA books – American Psychological Association (2010) Publication manual of the American Psychological association (6th ed.) Washington, DC.

API - Martins D, Gemmill B, Eardley C, Kinuthia W, Kwapong P, Gordon I (2003) The Plan of Action of the African Pollinator Initiative. *African Pollinator Initiave*: 1–41.
Retrieved from www.fao.org/docrep/010/a1490e/a1490e00.htm

Araújo ED, Costa M, Chaud-Neto J, Fowler HG (2004) Body size and flight distance in stingless bees (Hymenoptera:Meliponini): inference of flight range and possible ecological implications. *Brazilian Journal of Biology* 64(3B):563-568.

Araújo VA, Antonini Y, Araújo APA (2006) Diversity of Bees and their Floral Resources at Altitudinal Areas in the Southern Espinhaço Range, Minas Gerais, Brazil. *Neotropical Entomology* 35(1):030-040.

Araújo MB, Ferri-Yánez F, Bozinovic F, Marquet PA, Valladares F, Chown SL (2013) Heat freezes niche evolution. *Ecology Letters* 16: 1206–1219.

Arroyo MTK, Primack RB, Armesto J (1982) Community studies in pollination ecology in the high temperate Andes of central Chile. I. Pollination mechanisms and altitudinal variation. *American Journal of Botany* 69: 82-97.

Ascher J, Pickering J (2018, November 1) World Bee Diversity. Retrieved from https://www.discoverlife.org/nh/cl/counts/Apoidea_species.html

Atkinson D (1994) Temperature and organism size: a biological law for ectotherms? *Advances in Ecological Research* 25: 1–58.

Bagchi R, Gallery RE, Gripenberg S, Gurr SJ, Narayan L, Addis CE *et al.* (2014) Pathogens and insect herbivores drive rainforest plant diversity and composition. *Nature* 506: 85–88.

Baptista N, António T, Branch WR (2018) Amphibians and reptiles of the Tundavala region of the Angolan Escarpment. In: Climate change and adaptive land management in southern Africa – assessments, changes, challenges, and solutions (ed. by Revermann R, Krewenka KM, Schmiedel U, Olwoch JM, Helmschrot J, Jürgens N). *Biodiversity & Ecology* 6, Klaus Hess Publishers, Göttingen & Windhoek: 397-403.

Banaszak J, Banaszak-Cibicka W, Szefer P (2014) Guidelines on sampling intensity of bees (Hymenoptera: Apoidea: Apiformes). Journal of Insect Conservation 18: 651-656.

Baptista NL, Mills MSL (2018) Angola white-headed barbet *Lybius [leucocephalus] leucogaster* rediscovered. *Bull ABC* 25 (2).

Basu P, Parui Ak, Chatterjee S, Dutta A, Chakraborty P, Roberts S, Smith B (2016) Scale dependent drivers of wild bee diversity in tropical heterogeneous agricultural landscapes. *Ecology and Evolution* 6(19): 6983-6992.

Batley M, Hogendoorn K (2009) Diversity and conservation status of native Australian bees. *Apidologie* 40: 347–354.

Beck J, Altermatt F, Hagmann R, Lang S (2010) Seasonality in the altitude-diversity pattern of Alpine moths. *Basic and applied ecology* 11(8): 714–722.

Bendifallah L, Louadi K, Doumandji S (2013) Bee fauna potential visitors of Coriander flowers *Coriandrum sativum* L. (Apiaceae) in the Mitidja Area (Algeria). *Journal of Apicultural Science* 57(2): 59–70.

Berner D, Kirner C, Blanckenhorn WU (2004) Grasshopper Populations across 2000 m of Altitude : Is There Life History Adaptation? *Ecography* 27(6): 733–740.

Bersacola E, Svensson MS, Bearder SK (2015) Niche Partitioning and Environmental Factors Affecting Abundance of Strepsirrhines in Angola. *American Journal of Primatology* 77: 1179–1192.

Bidau CJ, Marti DA (2007) Clinal variation of body size in *Dichroplus pratensis* (Orthoptera: Acrididae): Inversion of Bergmann's and Rensch's rules. *Annals of the Entomological Society of America* 100(6): 850-860.

Biesmeijer JC, Roberts SPM, Reemer M, Ohlemuller R, Edwards M, Peeters T, Schaffers AP, Potts SG, Kleukers R, Thomas CD, Settele J, Kunin WE (2006) Parallel declines in pollinators and insect pollinated plants in Britain and the Netherlands. *Science* 313: 351–354.

Bishop JA, Armbruster WS (1999) Thermoregulatory abilities of Alaskan bees: effects of size, phylogeny and ecology. *Functional Ecology* 13: 711–724.

Blackburn TM, Gaston KJ, Loder N (1999) Geographic gradients in body size: a clarification of Bergmann's rule. *Diversity and Distributions* 5: 165–174.

Bosch J, Vicens N (2006) Relationship between body size, provisioning rate, longevity and reproductive success in females of the solitary bee *Osmia cornuta*. *Behavioral Ecology and Sociobiology* 60: 26-33.

Braat LC, de Groot R (2012) The ecosystem services agenda: bridging the worlds of natural science and economics, conservation and development, and public and private policy. *Ecosystem services* 1(1): 4–15.

Brehm G, Fiedler K (2004) Bergmann's rule does not apply to geometrid moths along an elevational gradient in an Andean montane rain forest. *Global ecology and biogeography* 13: 7-14.

Brehm G, Colwell RK, Kluge J (2007) The role of environment and mid-domain effect on moth species richness along a tropical elevational gradient. *Global Ecology and Biogeography* 16: 205–219.

Brosi BJ, Daily GC, Ehrlich PR (2007) Bee community shifts with landscape context in a tropical countryside. *Ecological Applications* 17(2): 418-430.

Brown M, Paxton R (2009) The conservation of bees: a global perspective. *Apidologie* 40(3): 410–416.

Buchmann SL (1987) The ecology of oil flowers and their bees. *Annual Review of Ecology and Systematics* 18: 343-369.

Buckley LB, Kingsolver JG (2012) Functional and phylogenetic approaches to forecasting species' responses to climate change. *Annual Review of Ecology, Evolution, and Systematics* 43: 205–226.

Burkle L, Marlin J, Knight T (2013) Plant-Pollinator Interactions over 120 Years: Loss of Species, Co-Occurence, and Function. *Science (March)*: 1611–1616.

Cáceres A, Melo M, Barlow J, Cardoso P, Maiato F, Mills MSL (2014) Threatened birds of the Angolan Central Escarpment: distribution and response to habitat change at Kumbira Forest. Fauna & Flora International, *Oryx*: 1-8.

Cáceres A, Melo M, Barlow J, Mills MSL (2016) Radiotelemetry reveals key data for the conservation of *Sheppardia gabela* (Rand, 1957) in the Angolan Escarpment forest. *African Journal of Ecology* 54: 317-327.

Cáceres A, Melo M, Barlow J, de Lima RF, Mills MSL (2017) Drivers of bird diversity in an understudied African centre of endemism: The Angolan Central Escarpment Forest. *Bird Conservation International* 27: 256-268.

Cane JH (1987) Estimation of bee size using intertegular span (Apoidea). Journal of the Kansas Entomological Society 60: 145-147.

Carvalheiro LG, Seymour CL, Veldtman R, Nicolson SW (2010) pollination services decline with distance from natural habitat even in biodiversity-rich areas. *Journal of Applied Ecology* 47: 810-820.

Carvalheiro LG, Veldtman R, Shenkute AG, Tesfay GB, Pirk CWW, Donaldson JS *et al.* (2011) Natural and within-farmland biodiversity enhances crop productivity. *Ecology Letters* 14: 251-259.

Carvalheiro LG, Seymour CL, Nicolson SW, Veldtman R (2012) Creating patches of native flowers facilitates crop pollination in large agricultural fields: mango as a case study. *Journal of Applied Ecology* 49: 1373-1383.

Castelo C (2012) Scientific research and Portuguese colonial policy: developments and articulations, 1936-1974. História, Ciências, Saúde – *Manguinhos* 19(2): 1–18.

Ceríaco L, Bauer A, Blackburn D, Lavres A (2014) The herpetofauna of the Capanda Dam region, Malanje, Angola. *Herpetological Review* 45(4): 667-674.

Ceríaco LMP, Marques MP, Bandeira SA *et al.* (2016) Anfíbios e Répteis do Parque Nacional da Cangandala. Instituto Nacional da Biodiversidade e Áreas de Conservação/Museu Nacional de História Natural e da Ciência, Luanda/Lisboa, 96 pp.

Chen I-C, Hill JK, Ohlemuller R, Roy DB, Thomas CD (2011) Rapid Range Shifts of Species Associated with High Levels of Climate Warming. *Science*, New Series 333 (6045): 1024-1026.

Chiawo DO, Ogol CKPO, Kioko EN, Otiende VA, Gikungu MW (2017) Bee Diversity and Floral Resources Along a Disturbance Gradient in Kaya Muhaka Forest and Surrounding Farmlands of Coastal Kenya. *Journal of Pollination Ecology* 20(6): 51–59.

Chown SL, Klok CJ (2003) Altitudinal Body Size Clines: Latitudinal Effects Associated with Changing Seasonality. *Ecography* 26(4): 445–455.

Chown SL, Gaston KJ (2010) Body size variation in insects: a macroecological perspective. *Biological Reviews* 85: 139–169.

Clark VR, Barker NP, Mucina L (2011) The Great Escarpment of southern Africa: a new frontier for biodiversity exploration. *Biodiversity Conservation* 20: 2543–2561.

Clarke KR, Warwick RM (2001) Change in marine communities: an approach to statistical analysis and interpretation, 2nd edition. *Primer-E Ltd*. Plymouth Marine Laboratory, UK.

Classen A, Peters MK, Kindeketa WJ, Appelhans T, Eardley CD, Gikungu MW *et al.* (2015) Temperature versus resource constraints: Which factors determine bee diversity on Mount Kilimanjaro, Tanzania? *Global Ecology and Biogeography* 24(6): 642–652.

Classen A, Steffan-Dewenter I, Kindeketa WJ, Peters MK (2017) Integrating intraspecific variation in community ecology unifies theories on body size shifts along climatic gradients. *Functional Ecology* 31: 768-777.

Clausnitzer V, Koch R, Dijkstra K-DB, Boudot J-P, Kipping J, Samraoui B *et al.* (2012) Focus on African freshwaters: hotspots of dragonfly diversity and conservation concern. *Frontiers in Ecology and the Environment* 10: 129–134.

Coaton WGH, Sheasby JL (1972) Preliminary report on a survey of the termites (Isoptera) of South West Africa. Cimbebasia Memoir No. 1.

Coddington JA, Agnarsson I, Miller JA, Kuntner M, Hormiga G (2009) Undersampling bias: the null hypothesis for singleton species in tropical arthropod surveys. *Journal of Animal Ecology* 78: 573-584.

Coelho MS, Carneiro MAA, Branco CA, Borges RAX, Fernandes GW (2017) Galling insects of the Brazilian Páramos: Species richness and composition along high-altitude grasslands. *Environmental Entomology* 46(6): 1–11.

Colville JF, Potts AJ, Bradshaw PL, Measey GJ, Snijman D, Picker MD *et al.* (2014) Floristic and faunal Cape biochoria: do they exist?. In: Fynbos: Ecology, evolution and conservation of a megadiverse region. Allsopp N, Colville JF, Verboom GA (eds). Oxford University Press chapter 4: 73-92.

Colwell RK (2013) EstimateS: statistical estimation of species richness and shared species from samples. Version 9.1.0. (www.viceroy.eeb.uconn.edu/estimates).

Costa FV, Mello R, Lana TC, Neves FS (2015) Ant fauna in megadiverse mountains: a checklist for the rocky grasslands. *Sociobiology* 62(2): 228-245.

Crawford-Cabral J, Mesquitela LM (1989) Índice toponímico de colheitas zoológicas em Angola. *Instituto de Investigação Científica Tropical. Estudos, Ensaios e Documentos* 151: 206pp.

Cronin AL (2001) Social Flexibility in a Primitively Social Allodapine Bee (Hymenoptera: Apidae): Results of a Translocation Experiment. *Oikos* 94(2): 337–343.

Cunningham SA (2000) Depressed pollination in habitat fragments causes low fruit set. *Proceedings of the royal society of London B: Biological sciences* 267: 1149–1152.

Cvetkovic' D, Tomas˘evic' N, Ficetola GF, Crnobrnja-Isailovic' J, Miaud C (2008) Bergmann's rule in amphibians: combining demographic and ecological parameters to explain body size variation among populations in the common toad *Bufo bufo*. *Journal of Zoological Systematics and Evolutionary Research* 47: 171-80

Damuth J (1987) Interspecific allometry of population density in mammals and other animals: the independence of body mass and population energy-use. *Biological Journal of the Linnean Society* 31(3): 193-246.

Danforth BN, Sipes S, Fang J, Brady SG (2006) The history of early bee diversification based on five genes plus morphology. *Proceedings of the national academy of sciences* 103(41): 15118–15123.

Davidowitz G, D'Amico LJ, Nijhout HF (2004) The effects of environmental variation on a mechanism that controls insect body size. *Evolutionary Ecology Research* 6: 49-62.

Dean WRJ, Melo M, Mills MSL (2019) The avifauna of Angola: Richness, endemism and rarity. In: Huntley BJ, Russo V, Lages F, Ferrand N (eds) Biodiversity of Angola. Science & conservation: a modern synthesis. Springer, Berlin.

Devoto, M, Medan, D, Montaldo, NH (2005) Patterns of interaction between plants and pollinators along an environmental gradient. *Oikos* 109: 461-472.

Diamond SE, Frame AM, Martin RA, Buckley LB (2011) Species' traits predict phenological responses to climate change in butterflies. *Ecology* 92: 1005–1012.

Dietemann V, Lubbe A, Crewe RM (2006) Human factors facilitating the spread of a parasitic honey bee in South Africa. *Journal of Economic Entomology* 99(November): 7–13.

Diniz AC (2006) Características Mesológicas de Angola – Descrição e correlação dos aspectos fisiográficos dos solos e da vegetação das zonas agrícolas angolanas. Zona agrícola 15. *Instituto Português de Apoio ao Desenvolvimento*. 2ª Edição. Lisboa.

Dunne JA, Saleska SR, Fischer ML, Harte J (2004) Integrating experimental and gradient methods in ecological climate change research. *Ecology* 85: 904–916.

Eardley C (2002) Afrotropical Bees Now: What Next? In: Kevan P, Imperatriz-Fonseca VL (eds) - Pollinating Bees – The Conservation Link Between Agriculture and Nature - Ministry of Environment / Brasília: p.97-104.

Eardley CD, Gikungu M, Schwarz MP (2009) Bee conservation in Sub-Saharan Africa and Madagascar: diversity, status and threats. *Apidologie* 40(3): 355–366.

Eardley CD, Urban R (2010) Catalogue of Afrotropical bees (Hymenoptera: Apoidea: Apiformes). *Zootaxa* 2455: 1–548.

Eardley C. 2013. A taxonomic revision of the southern African leaf-cutter bees, *Megachile* Latreille *sensu stricto* and *Heriadopsis* Cockerell (Hymenoptera: Apoidea: Megachilidae). *Zootaxa* 3601 (1): 001–133

Eilers EJ, Kremen C, Greenleaf SS, Garber AK, Klein AM (2011) Contribution of pollinator-mediated crops to nutrients in the human food supply. *PLoS ONE* 6(6): e21363.

Elizalde D (2018, December) Strengthening the institutional network in Angola to mobilize biodiversity data. Retrieved from https://www.gbif.org/pt/project/82775/strengthening-the-institutional-network-in-angola-to-mobilize-biodiversity-data.

Elizalde D, Elizalde S, Lutondo E, Groom R, Kersch K (2019) Luando Natural Integral Reserve, Angola - A large and medium sized mammals survey. Instituto Nacional da Biodiversidade e Áreas de Conservação - Range Wide Conservation Program for Cheetah and African Wild Dogs, Luanda. Unpublished report.

Environment Ministry (1999) The São Paulo Declaration on Pollinators. Brasilia - Brazil.

Fernandes GW, Almeida HA, Nunes CA, Xavier JHA, Cobb NS, Carneiro MAA, *et al.* (2016) Cerrado to Rupestrian Grasslands: Patterns of species distribution and the forces shaping them along an altitudinal gradient. In: Fernandes GW, editor. *Ecology and Conservation of Mountaintop Grasslands in Brazil.* Switzerland: Springer International Publishing: 345-378.

Figueiredo E, Smith GF, César J (2009) The Flora of Angola: First Record of Diversity and Endemism The flora of Angola: first record of diversity and endemism. *Taxon 58*(1): 233–236.

Freitas BM, Imperatriz-Fonseca VL, Medina LM, Kleinert A de MP, Galetto L, Nates-Parra G *et al.* (2009) Diversity, threats and conservation of native bees in the Neotropics. *Apidologie* 40(3): 332–346.

Fu C, Wu J, Wang X, Lei G, Chen J (2004) Patterns of diversity, altitudinal range and body size among freshwater fishes in the Yangtze River basin, China. *Global Ecology and Biogeography* 13: 543-552.
Funston PJ, Henschel P, Petracca L, Maclennan S, Whitesell C, Fabiano E, Castro I (2017) The distribution and status of lion and other large carnivores in Luengue-Luiana and Mavinga National Parks. Angola. KAZA TFCA Secretariat (KAZA).

Garibaldi LA, Aizen MA, Klein AM, Cunningham SA, Harder LD (2011a) Global growth and stability of agricultural yield decrease with pollinator dependence. *PNAS* 108(14): 5909-5914.

Gemmil-Herren B, Ochieng AO (2008) Role of native bees and natural habitats in eggplant (*Solanun melonguera*) pollination in Kenya. *Agriculture, Ecosystems and Environment* 127: 31-36.

Gess SK, Gess FW (2014) Wasps and bees of southern Africa. SANBI biodiversity Series.

Ghalambor CK, Huey RB, Martin PR, Tewksbury JJ, Wang G (2006) Are mountain passes higher in the tropics? Janzen's hypothesis revisited. *Integrated and Comparative Biology* 46(1): 5-17.

Gikungu MW (2006) Bee Diversity and some Aspects of their Ecological Interactions with Plants in a Successional Tropical Community. PhD book, University of Bonn.

Gikungu M, Wittmann D, Irungu D, Kraemer M (2011) Bee diversity along a forest regeneration gradient in Western Kenya. *Journal of Apicultural Research* 50(1): 22–34.

Gillman LN, McCowan LSC, Wright SD (2012) The tempo of genetic evolution in birds: body mass and climate effects. *Journal of Biogeography* 39: 1567–1572

Giovanetti M, Lasso E (2005) Body size, loading capacity and rate of reproduction in the communal bee *Andrena agilissima* (Hymenoptera: Andrenidae) *Apidologie* 36: 439-447

Godfray HCJ, Beddington JR, Crute IR, Haddad L, Lawrence D, Muir JF et al. (2010) Food security: the challenge of feeding 9 billion people. *Science* 327: 812-818.

Golder Associates (2006) Biodiversity and Social Scan of Lunda Norte, Angola. Relatório de consultoria para a De Beers

Angola Prospecting Limited, Luanda. 147 pp.

Gonçalves FMP, Goyder DJ (2016) A brief botanical survey into Kimbira forest, an isolated patch of Guineo-Congolian biome. *Phytokeys* 65: 1-14.

Gonçalves FMP, Tchamba J, Goyder DJ (2016) *Schistostephium crataegifolium* (Compositae: Anthemideae), a new generic record for Angola. *Bothalia* 46(1): a2029.

Goodman SM, Benstead JP, Schutz H (2003) The Natural History of Madagascar. *Chicago, IL: University of Chicago Press.*

Gotelli NJ, Colwell RK (2001) Quantifying biodiversity: procedures and pitfalls in the measurement and comparison of species richness. *Ecology Letters* 4: 379–391.

Gotelli NJ, Chao A (2013) Measuring and estimating species richness, species diversity, and biotic similarity from sampling data. *Encyclopedia of Biodiversity* vol. 5: 195-211

Goulson D, Nicholls E, Botías C, Rotheray EL (2015) Combined stress from parasites, pesticides and lack of flowers drives bee declines. *Science* 347, 1255957.

Goyder DJ, Gonçalves FMP (2019) The flora of Angola: collectors, richness and endemism. In: Huntley BJ, Russo V, Lages F, Ferrand N (eds) Biodiversity of Angola. Science & conservation: a modern synthesis. Springer, Berlin.

Grace A (2010) Introductory biogeography to bees of the eastern Mediterranean and near east. Bexhill Museum. Sussex. United Kingdom.

Graham CH, Ferrier S, Huettman F, Moritz C, Peterson AT (2004) New developments in museum-based informatics and applications in biodiversity analysis. *Trends in Ecology and Evolution 19*(9): 497-503.

Grandvaux-Barbosa LA (1970) Carta fitogeográfica de Angola. Instituto de Investigação Científica de Angola, Luanda. 323 p.

Greenleaf SS, Kremen C (2006a) Wild bee species increase tomato production and respond differently to surrounding land use in northern California. *Biological Conservation* 133: 81-87.

Greenleaf SS, Kremen C (2006b) Wild bees enhance honey bees' pollination of hybrid sunflower. PNAS 103(37): 13890-13895.

Greenleaf SS, Williams NM, Winfree R, Kremen C (2007) Bee foraging ranges and their relationship to body size. *Oecologia* 153: 589-596.

Groom R, Elizalde S, Elizalde D, de Sá S, Alexandre G (2018) Quiçama National Park, Angola - Large and medium sized terrestrial mammals survey. Instituto Nacional da Biodiversidade e Áreas de Conservação - Range Wide Conservation Program for Cheetah and African Wild Dogs, Luanda. Unpublished report.

Grytnes J-A, McCain CM (2007) Elevational trends in biodiversity. In Levin SA(ed.). *Encyclopedia of Biodiversity. Elsevier*, New York: 1–8.

Guimarães F (1992) The Origins of the Angolan Civil War: International politics and domestic political conflict 1961-1976. PhD book, The London School of Economics and Political Science.

Guiraud M, Chagnoux S (2018, November 17) MNHN – Museum national d'Histoire naturelle – Entomology. Retrieved from: https://www.gbif.org/dataset/2cc23bac-e94b-414f-95ab-cc838c03f765#description

Hall BP (1960) The faunistic importance of the scarp of Angola. *Ibis* 102: 420–442.

Hansen MC, Stehman SV, Potapov PV (2010) Quantification of global gross forest cover loss. *PNAS* 107(19): 8650-8655.

Hansen MC, Potapov PV, Moore R, Hancher M, Turubanova SA, Tyukavina A, *et al.* (2013) High-Resolution Global Maps of 21st-Century Forest Cover Change. *Science* 342 (15 November): 850–53. Data available on-line from: http://earthenginepartners.appspot.com/science-2013-global-forest.

Hawkins BA, Field R, Cornell HV, Currie DJ, Guégan JF, Kaufman DM, *et al.* (2003) Energy, water, and broad-scale geographic patterns of species richness. *Ecology* 84: 3105–3117.

Hepburn HR, Guye SG (1993) An annotated bibliography of the Cape honeybee, *Apis mellifera capensis* Eschscholtz (Hymenoptera: Apidae). *African Entomology* 1(2): 235–252.

Hillebrand H (2004) On the generality of the latitudinal diversity gradient. *The American Naturaliste* 163(2): 192–211.
Hind DJN, Goyder DJ (2014) *Stomatanthes tundavalaensis* (Compositae: Eupatorieae: Eupatoriinae), a new species from Huíla Province, Angola, and a synopsis of the African species of Stomatanthes. *Kew Bulletin* 69(4): 9545.

Hines CJH (2018a) The avifauna of Bicuar National Park. Unpublished report. INBAC-RWCP

Hines CJH (2018b) Vegetation of Bicuar National Park: a rapid survey (28/03 – 05/04/2018). Unpublished report. INBAC-RWCP.

Hodkinson ID (2005) Terrestrial insects along elevation gradients: species and community responses to altitude. *Biological reviews* 80(3): 489.

Hoeksema BW, van der Land, van der Meij SET, van der Ofwegen LP, Reijnen BT, van Soest RWM *et al.* (2011) Unforeseen importance of historical collections as baselines to determine biotic change of coral reefs: the Saba Bank. *Marine Ecology* 32: 135–141.

Hoiss B, Krauss J, Potts SG, Roberts S, Steffan-Dewenter I (2012) Altitude acts as an environmental filter on phylogenetic composition, traits and diversity in bee communities. *Proceedings of the royal society B: Biological sciences* 279(1746): 4447–4456.

Hoiss B, Gaviria J, Leingärtner A, Krauss J, Steffan-Dewenter I (2013) Combined effects of climate and management on plant diversity and pollination type in alpine grasslands. *Diversity and Distributions* 19: 386–395.

Holland PG, Steyn DG (1975) Vegetational responses to latitudinal in slope angle and aspect. *Journal of Biogeography* 2(3): 179-183.

Hortal J, Borges PAV, Gaspar C (2006) Evaluating the performance of species richness estimators: sensitivity to grain size. *Journal of Animal Ecology* 75: 274-287.

Hozumi S, Mateus S, Kudô K, Kuwahara T, Yamane S, Zucchi R (2010) Nest thermoregulation in *Polybia scutellaris* (White) (Hymenoptera: Vespidae). *Neotropical Entomology* 39(5): 826-828.

Huber BA, Sinclair BJ, Lampe KH (2005) African Biodiversity: Molecules, Organisms, Ecosystems. *Springer,* Netherlands: pp. 172

Huntley BJ (1974) Outlines of wildlife conservation in Angola. *Journal of the Southern African Wildlife Management Association* 4: 157–166.

Hurlbert AH (2004) Species–energy relationships and habitat complexity in bird communities. *Ecology Letters* 7: 714–720.

van Jaarsveld EJ (2010) Angola Botanical Expedition: Succulent Treasures, January 2009. *Aloe* 47 (1)

Jennersten O (1988) Pollination in *Dianthus deltoides* (Caryophyllaceae): Effects of habitat fragmentation on visitation and seed set. *Conservation biology* 2(4): 359–366.

Jiménez-Valverde A, Lobo JM (2005) Determining a combined sampling procedure for a reliable estimation of Araneidae and Thomisidae assemblages (Arachnida, Araneae). The Journal of Arachnology 33: 33-42.

Johannsmeier M, Swart D, Morudu T (1997) Honeybees in an avocado orchard: Forager distribution, influence on fruit set and colony development. South African Avocado Growers' Association Yearbook 20: 39–41.

Johnson MA, Chiang RA (2015) Quantitative guidance for stove usage and performance to achieve health and environmental targets. *Environmental Health Perspectives* 123(8): 820-826.

Karanja R, Njoroge GN, Gikungu M, Newton LE (2010) Bee interactions with wild flora around organic and conventional coffee farms in Kiambu district, central Kenya. *Journal of Pollination Ecology* 2(2): 7–12.

Karunaratne WAIP, Edirisinghe JP (2008) Diversity of bees at different altitudes in the Knuckles Forest Reserve. *Ceylon Journal of Sciences (Biological Sciences)* 37(1): 61-72.

Kearns C, Inouye D, Waser N (1998) Endangered mutualisms: The conservation of plant-pollinator interactions. *Annual review of ecology and systematics* 29: 83–112.

Keating BA, Carberry PS, Bindraban PS, Asseng S, Meinke H, Dixon J (2010) Eco-efficient agriculture: conepts, challenges and opportunities. *Crop Sciences* 50: 109–119.

Kellerman V, Overgaard J, Hoffmann AA, Flojgaard C, Svenning J-C, Loeschcke V (2012) Upper thermal limits of *Drosophila* are linked to species distributions and strongly constrained phylogenetically. *Proceedings of the National Academy of Sciences* U.S.A. 109: 16228–16233.

Kerr W (1984) Virgílio de Portugal Brito Araújo (1919-1983) - obituary. Manaus. *Acta Amazonica* 14(1-2).

Kerr JT, Pindar A, Galpern P, Packer L, Potts SG, Roberts SM *et al.* (2015) Climate change impacts on bumblebees converge across continents. *ScienceMag* 349(6244): 177-180.

Kingsolver JG, Huey RB (2008) Size, temperature, and fitness: three rules. *Evolutionary Ecology Research* 10: 251-268.

Kingsolver JG, Woods HA, Buckley LB, Potter KA, Mclean HJ, Higgins JK (2011) Complex life cycles and the responses of insects to climate change. *Integrative and Comparative Biology* 51(5): 719–732.

Kipping J, Clausnitzer V, Elizalde SRFF, Dijkstra KDB (2017) The dragonflies and damselflies (Odonata) of Angola. *African Invertebrates* 58(1): 65–91.

Kipping J, Clausnitzer V, Elizalde SRFF, Dijkstra KDB (2019) The dragonflies and damselflies (Odonata) of Angola: an updated synthesis. In: Huntley BJ, Russo V, Lages F, Ferrand N (eds) Biodiversity of Angola. Science & conservation: a modern synthesis. Springer, Berlin.

Klatt BJ, Holzschuh A, Westphal C, Clough Y, Smit I, Pawelzik E, Tscharntke T (2014) Bee pollination improves crop quality, shelf life and commercial value. *Proceedings of the Royal Society B: Biological Sciences* 281(1775): 20132440.

Klein A-M, Steffan-Dewenter I, Tscharntke T (2003) Pollination of *Coffea canephora* in relation to local and regional agroforestry management. *Journal of Applied Ecology* 40: 837-845.

Klein A-M, Vaissiere BE, Cane JH, Steffan-Dewenter I, Cunningham SA, Kremen C *et al.* (2007) Importance of pollinators in changing landscapes for world crops. *Proceedings of the royal society: Biological sciences*, 274(1608): 303–313.

Kleijn D, Winfree R, Bartomeus I, Carvalheiro LG, Henry M, Isaacs R *et al.* (2015) Delivery of crop pollination services is an insufficient argument for wild pollinator conservation. *Nature Communications* 6: 7414.

Kluser S, Peduzzi P (2007) Global pollinator decline: A literature review. UNEP/GRID- Europe.

Kocher SD, Pellissier L, Veller C, Purcell J, Nowak MA, Chapuisat M, *et al.* (2014) Transitions in social complexity along elevational gradients reveal a combined impact of season length and development time on social evolution. *Proceedings of the royal society: Biological sciences* 281: 20140627.

Koh I, Lonsdorf E, Williams N, Brittain C, Isaacs R, Gibbs J, Ricketts T (2016) Modeling the status, trends, and impacts of wild bee abundance in the United States. *PNAS* 113(1): 140–145.

Körner, C. (2000) Why are there global gradients in species richness? Mountains might hold the answer. *Trends in Ecology & Evolution* 15: 513–514.

Körner C (2007) The use of "altitude" in ecological research. *Trends in Ecology and Evolution* 22(11): 569–574.

Koski MH, Ashman T-L (2015) An altitudinal cline in UV floral pattern corresponds with a behavioural change of generalist pollinator assemblage. *Ecology* vol. 96(12): 3343-3353.

Kovac H, Stabentheiner A (2011) Thermoregulation of foraging honeybees on flowering plants. *Ecological Entomology* 36: 686–699.

Kremen C, Williams NM, Thorp RW (2002) Crop pollination from native bees at risk from agricultural intensification. *PNAS* 99(26): 16812-16816.

Kremen C, Williams N, Bugg R, Fay J, Thorp R (2004) The area requirements of an ecosystem service: crop pollination by native bee communities in California. 1109–1119.

Krishtalka L, Humphrey P (2000) Can Natural History Museums Capture the Future? *BioScience 50*(7): 611–617.

Kuedikuenda S, Xavier M (2009) Framework report on Angola's biodiversity. Ministry of Environment - Luanda.

Kuhlmann M (2005) Diversity, distribution patterns and endemism of Southern African Bees (Hymenoptera:Apoidea). *African Biodiversity* (July): 173–179.

Kuhlmann M (2009) Patterns of diversity, endemism and distribution of bees (Insecta: Hymenoptera : Anthophila) in southern Africa. *South African Journal of Botany* 75: 726–738.

Kumar A, Longino JT, Colwell RK, O'Donnell S (2007) Elevational patterns of diversity and abundance of eusocial paper wasps (Vespidae) in Costa Rica. *Biotropica* 5(3): 338–346.

Larsen FW, Bladt J, Rahbek C (2009) Indicator taxa revisited: useful for conservation planning? *Diversity and Distributions* 15:70–79.

Leite A, Cáceres A, Melo M, Mills MSL, Monteiro AT (2018) Reducing emissions from deforestation and forest degradation in Angola: Insights from the scarp forest conservation 'hotspot'. John Wiley & Sons, Ltd. *Land Degradation & Development* 29: 4291–4300.

Liao W, Lu X (2012) Adult body size = f (initial size + growth rate x age): explaining the proximate cause of Bergman's cline in a toad along altitudinal gradients. *Evolutionary Ecology* 26: 579-590.

Linder HP, de Klerk HM, Born J, Burgess ND, Fjeldså J, Rahbek C (2012) The partitioning of Africa: Statistically defined biogeographical regions in sub-Saharan Africa. *Journal of Biogeography* 39(7): 1189–1205.

Lomolino MV (2001) Elevation gradients of species-density: historical and prospective views. *Global Ecology & Biogeography* 10: 3-13.

Louadi K, Benachour K, Berchi S (2007) Floral visitation patterns of bees during spring in Constantine, Algeria. *African Entomology* 15(1): 209–213.

Lovett JC (2018) Angola: policy piece. *John Wiley & Sons* Ltd, African Journal of Ecology 56: 1–2.

Lundberg J, Moberg F (2003) Mobile link organisms and ecosystem functioning: implications for ecosystem resilience and management. *Ecosystems* 6: 87-98.

Maad J, Armbruster WS, Fenster CB (2013) Floral size variation in *Campanula rotundifolia* (Campanulaceae) along altitudinal gradients: patterns of possible selective mechanisms. *Nordic Journal of Botany* 31: 361-371.

Malo JE, Baonza J (2002) Are there predictable clines in plant-pollinator interactions along altitudinal gradients? The example of *Cytisus scoparius* (L.) Link in the Sierra de Guadarrama (Central Spain). Diversity and Distributions 8: 365-371.

Marques MP, Ceríaco LMP, Blackburn DC, Bauer AM (2018) Diversity and distribution of the amphibians and terrestrial reptiles of Angola. Atlas of historical and bibliographic records (1840–2017). *Proceedings of the California Academy of Sciences Series* 4(65): 1–501.

Martins KT, Gonzalez A, Lechowicz MJ (2015) Pollination services are mediated by bee functional diversity and landscape context. *Agriculture, Ecosystems and Environment* 200: 12-20.

Mason NWH, Lanoiselee C, Mouillot D, Irz P, Argillier C (2007) Functional characters combined with null models reveal inconsistency in mechanisms of species turnover in lacustrine fish communities. *Oecologia* 153: 441–452.

McGill BJ, Etienne RS, Gray JS, Alonso D, Anderson MJ, Benecha HK *et al.* (2007) Species abundance distributions: moving beyond single prediction theories to integration within an ecological framework. *Ecology Letters* 10: 995-1015.

Meiri S, Dayan T (2003) On the validity of Bergmann's rule. *Journal of Biogeography* 30: 331–351.

Meixner MD, Leta MA, Koeniger N, Fuchs S (2011) The honey bees of Ethiopia represent a new subspecies of *Apis mellifera – Apis mellifera simensis* n.ssp. *Apidologie* 42: 425-437.

Melin A, Rouget M, Midgley JJ, Donaldson JS (2014) Pollination ecosystem services in South African agricultural systems. *South African Journal of Science* 110(11/12): 1-9.

Mendes LF, Bívar de Sousa A, Figueira R, Serrano ARM (2013) Gazetteer of the Angolan localities known for beetles (Coleoptera) and butterflies (Lepidoptera: Papilionoidea). *Boletim da Sociedade Portuguesa de Entomologia* 228 (VIII-14): 258-292.

Mendes LF, Bivar-de-Sousa A, William MC (2019) The butterflies and skippers (Lepidoptera: Papilionoidea) of Angola: An updated checklist. In: Huntley BJ, Russo V, Lages F, Ferrand N (eds) *Biodiversity of Angola*. Science & conservation: a modern synthesis. Springer, Berlin.

Michener CD (1979) Biogeography of the bees. *Annals of the Missouri Botanical Garden* 66(3): 277–347.

Michener CD (2007) The bees of the world. *The Johns Hopkins University Press* (Vol. 85).

Milet-Pinheiro P, Ayasse M, Dobson HEM, Schlindwein C, Francke W, Dotterl S (2013) The chemical basis of host-plant recognition in a specialized bee pollinator. *Journal of Chemical Ecology* 39: 1347-1360.

Miller SE, Rogo LM (2001) Challenges and opportunities in understanding and utilization of African insect diversity. *Cimbebasia* 17(August 2000): 197–218.

Mills MSL, Dean WRJ (2007) Notes on Angolan birds: new country records, range extensions and taxonomic questions. *Ostrich* 78(1): 55–63.

Mills MSL, Dowd AD (2007) First records of Lemon Dove *Aplopelia larvata* for Angola. *Bull ABC* 14(1).

Mills MSL, Vaz Pinto P, Dean WRJ (2008) The avifauna of Cangandala National Park, Angola. *Bulletin of African Bird Club* 15: 113–116.

Mills MSL (2010) Angola's central scarp forests: patterns of bird diversity and conservation threats. *Biodiversity and conservation* 19: 1883–1903.

Mills MSL, Pinto PV, Haber S (2012) Grey-striped Francolin *Pternistis griseostriatus*: specimens, distribution and morphometrics. *Bull ABC* 19(2).

Mills MSL, Melo M (2013) The checklist of the birds of Angola/A Lista das Aves de Angola. Associação Angolana para Aves e Natureza & Birds Angola, Luanda.

Mills MSL (2018) The special birds of Angola/As Aves Especiais de Angola. *Go-Away-Birding*, Cape Town.

Morimoto Y, Gikungu M, Maundu P (2004) Pollinators of the bottle gourd (*Lagenaria siceraria*) observed in Kenya. *International Journal of Tropical Insect Science* 24(1): 79–86.

Morris RJ, Sinclair FH, Burwell CJ (2015) Food web structure changes with elevation but not rainforest stratum. *Ecography* 38: 792–802.

Myers N, Mittermeier RA, Mittermeier CG, daFonseca GAB, Kent J (2000) Biodiversity hotspot for conservation priorities. *Nature* 403: 853-858.

Natural History Museum (2018, November 1) Dataset: Hymenoptera Apoidea Anthophila. Retrieved from: http://data.nhm.ac.uk/dataset/56e711e6-c847-4f99-915a-6894bb5c5dea/resource/05ff2255-c38a-40c9-b657-4ccb55ab2feb?view_id=203a0ae5-6a14-480a-a407-27eeb9373858&filters=collectionCode%3Abmnh%28e%29

Neumann P, Carreck NL (2010) Honey bee colony losses. *Journal of Apiculture Research* 49(1): 1-6.

Nicolson SW (2011) Bee food: the chemistry and nutritional value of nectar, pollen and mixtures of the two. *African Zoology* 46(2): 197-204.

Njoroge GN, Gemmill B, Bussmann R, Newton LE, Ngumi VW (2004) Pollination ecology of *Citrullus lanatus* at Yatta, Kenya. *International Journal of Tropical Insect Science* 24(1): 73–77.

Nogués-Bravo D, Araújo MB, Romdal T, Rahbek C (2008) Scale effects and human impact on the elevational species richness gradients. *Nature*: 453.

Novotny´ V, Basset Y (2000) Rare species in communities of tropical insect herbivores: pondering the mystery of singletons. *Oikos* 89: 564–572.

Nunes CA, Braga RF, Figueira JEC, Neves FdS, Fernandes GW (2016) Dung beetles along a tropical altitudinal gradient: environmental filtering on taxonomic and functional diversity. *PLoS ONE* 11(6): e0157442.

Okavango Wilderness Project (2017) Resultados iniciais de Biodiversidade decorrentes da exploração das áreas de captação dos rios Cuito, Cuanavale e Cuando, no centro e sudeste de Angola (Maio de 2015 - Dezembro de 2016). Unpublished report. National Geographic.

Oldroyd BP (2007) What's killing American honey bees? *Plos Biol*. 5:1195–1199

Ollerton J, Winfree R, Tarrant S (2011) How many flowering plants are pollinated by animals? *Oikos* 120: 321-326.

Olson DM, Dinerstein E (1998) The Global 200: a representation approach to conserving the Earth's most biologically valuable ecoregions. *Conservation Biology* 12: 502–515.

Olson DM, Dinerstein E, Wikramanayake ED, Burgess ND, Powell GVN, Underwood EC, Kassem KR (2001) Terrestrial Ecoregions of the World: A New Map of Life on Earth. *BioScience 51*(11): 933–938.

Osorio-Canadas S, Arnan X, Rodrigo A, Torné-Noguera A, Molowny R, Bosch J (2016) Body size phenology in a regional bee fauna: a temporal extension of Bergmann's rule. *Ecology Letters* 19: 1395-1402.

Overton J, Fernandes S, Elizalde D, Groom R, Funston P (2017) A large mammal survey of Bicuar and Mupa National Parks, Angola – with special emphasis on the presence and status of cheetah and African wild dogs. National Institute of Biodiversity and Conservation Areas in partnership with the Range Wide Conservation Program for Cheetah and African Wild Dogs.

Paritsis J, Aizen MA (2008) Effects of exotic conifer plantations on the biodiversity of understory plants, epigeal beetles and birds in *Nothofagus dombeyi* forests. *Forest Ecology and Management* 255: 1575–1583.

Patiny S, Michez D (2007) Biogeography of bees (Hymenoptera, Apoidea) in Sahara and the Arabian deserts. *Insect Systematics & Evolution* 38(1): 19–34.

Patiny S, Rasmont P, Michez D (2009) A survey and review of the status of wild bees in the West-Palaearctic region. *Apidologie* 40(3): 313–331.

Pauly A, Brooks RW, Nilsson A, Apesenko Y, Eardley CD, Terzo M, *et al.* (2001) Hymenoptera: Apoidea de Madagascar et des îles voisines. *Annales Sciences Zoologiques* 286: 412.

Pauly A. 2008. Catalogue of the sub-Saharan species of the genus *Seladonia* Robertson, 1918, with description of two new species (Hymenoptera: Apoidea: Halictidae). *Zool. Med. Leiden* 82: 391-400.

Pauly A. 2014. Les Abeilles des Graminées ou Lipotriches Gerstaecker, 1858, sensu stricto (Hymenoptera : Apoidea : Halictidae : Nomiinae) de l'Afrique subsaharienne. *Belgian Journal of Entomology*, 20 : 1-393.

Peel MC, Finlayson BL, Mcmahon TA (2007) Updated world map of the Koppen-Geiger climate classification. *Hydrology and earth system sciences discussions, European Geosciences Union* 4(2): 439–473.

Pereboom JJM, Biesmeijer JC (2003) Thermal constraints for stingless bee foragers: the importance of body size and coloration. *Oecologia* 137: 42-50.

Perillo LN, Neves FDS, Antonini Y, Martins RP (2017) Compositional changes in bee and wasp communities along Neotropical mountain altitudinal gradient. *PLoS ONE* 12(7): 1–14.

Peruquetti RC (2009) Variação do tamanho corporal de machos de Eulaema nigrita Lepeletier (Hymenoptera, Apidae, Euglossini). Resposta materna à flutuação de recursos? *Revista Brazileira de Zoologia* 20(2): 207-212.

Petchey OL, Gaston KJ (2006) Functional diversity: back to basics and looking forward. *Ecology Letters* 9: 741–758.

Peters MK, Hemp A, Appelhans T, Behler C, Classen A, Detsch F, *et al.* (2016) Predictors of elevational biodiversity gradients change from single taxa to the multi-taxa community level. *Nature communications* 7: 13736.

Petersen FT, Meier R (2003) Testing species-richness estimation methods on single-sample collection data using the Danish Diptera. *Biodiversity and Conservation* 12: 667-686.

Phalan B, Balmford A, Green RE, Scharlemann JPW (2011) Minimising harm to biodiversity of producing more food globally. *Food Policy* 36: S62–S71.

Picker MD (2018) Survey of insects of Bicuar National Park, with emphasis on butterflies. Biological Characterization of Bicuar National Park. Unpublished report. INBAC-RWCP

de Portugal Araújo V (1955) Notas sobre colónias de Meliponíneas de Angola – África. *Dusenia* VI: 3-4.

de Portugal Araújo V (1957) Colmeias e utensílios para a cultura de abelhas sem ferrão, com especial referência à espécie Cólo (*Trigona (Meliponula) bocandei* Spinola). *Gazeta Agrícola de Angola*: 513-517.

Potts SG, Petanidou T, Roberts S, O'Toole C, Hulbert A, Willmer P (2006) Plant pollinator biodiversity and pollination services in a complex Mediterranean landscape. *Biological Conservation* 129: 519–529.

Potts SG, Biesmeijer JC, Kremen C, Neumann P, Schweiger O, Kunin WE (2010) Global pollinator declines: Trends, impacts and drivers. *Trends in Ecology and Evolution* 25(6): 345–353.

Praz CJ, Müller A, Danforth BN, Griswold TL, Widmer A, Dorn S (2008) Phylogeny and biogeography of bees of the tribe Osmiini (Hymenoptera: Megachilidae). *Molecular phylogenetics and evolution* 49(1): 185–197.

Price TD, Hooper DM, Buchanan CD, Johansson US, Tietze DT, Alström P *et al.* (2014) Niche filling slows the diversification of Himalayan songbirds. *Nature* 509: 222–225.

Pyke GH, Ehrlich PR (2010) Biological collections and ecological/environmental research: a review, some observations and a look to the future. *Biological Reviews 85*(2): 247–266.

Rader R, Howlett BG, Cunningham SA, Westcott DA, Newstrom-Lloyd LE, Walker MK *et al.* (2009) Alternative pollinator taxa are equally efficient but not as effective as the honeybee in a mass flowering crop. *Journal of Applied Ecology* 46: 1080–1087.

Rader R, Howlett BG, Cunningham SA, Westcott DA, Edwards W (2012) Spatial and temporal variation in pollinator effectiveness: do unmanaged insects provide consistent pollination services to mass flowering crops? *Journal of Applied Ecology* 49(1): 126–134.

Rahbek, C (1995) The Elevational Gradient of Species Richness: A Uniform Pattern? *Ecography* 18(2): 200-205.

Rahbek, C (2005) The role of spatial scale and the perception of large-scale species-richness patterns. *Ecology Letters* 8: 224–239.

Ramirez SR, Roubik DW, Skov C, Pierce NE (2010) Phylogeny, diversification patterns and historical biogeography of euglossine orchid bees (Hymenoptera: Apidae). *Biological Journal of the Linnean Society* 100: 552-572.

Ribeiro S (1973, Outubro) A origem e história da companhia – O diamante, a Diamang e a Lunda. Retrieved from https://www.diamang.com.

Rodger JG, Balkwill K, Gemmill B (2004) African pollination studies: where are the gaps? *International Journal of Tropical Insect Science* 24(1): 5–28.

Rodrigues P, Figueira R, Vaz Pinto P, Araújo MB, Beja P (2015) A biogeographical regionalization of Angolan mammals. *Mammal Review* 45: 103-116.

Romeiras MM, Figueira R, Duarte MC, Beja P, Darbyshire I (2014) Documenting biogeographical patterns of African timber species using herbarium records: A conservation perspective based on native trees from Angola. *PLoS ONE* 9(7).

Ron T (2015) Preliminary Assessment of eight National Parks and one Strict Nature Reserve for planning further Project and Government Interventions. Ministério do Ambiente – *Projecto Nacional da Biodiversidade: Conservação do Parque Nacional do Iona.*

Rosário Nunes JF, Tordo GC (1960) Prospecções e ensaios experimentais apícolas em Angola. Estudos, Ensaios & Documentos 70. *Junta de Investigações do Ultramar* – Lisboa.

Roubik DW (2001) Ups and downs in pollinator populations when is there a decline?. *Conservation Ecology* 5(1).

Rourke BC (2000) Geographic and altitudinal variation in water balance and metabolic rate in a California grasshopper, *Melanoplus sanguinipes*. *Journal of experimental biology* 203: 2699–2712.

Ruttner F, Elmi MP, Fuchs S (2000) Ecoclines in the Near East along 36 degrees N latitude in *Apis mellifera* L. *Apidologie* 31: 157–165.

Ryan PG, Sinclair I, Cohen C, Mills MSL, Spottiswoode CN, Cassidy R (2004) The conservation status and vocalizations of threatened birds from the scarp forests of the Western Angola Endemic Bird Area. *Bird Conservation International* 14: 247–260.

Scaven VL, Rafferty NE (2013) Physiological effects of climate warming on flowering plants and insect pollinators and potential consequences for their interactions. *Current Zoology* 59(3): 418-426.

Schellenberger Costa D, Classen A, Ferger S, Helbig-Bonitz M, Peters M, Boêhning-Gaese K, *et al.* (2017) Relationships between abiotic environment, plant functional traits, and animal body size at Mount Kilimanjaro, Tanzania. *PLoS ONE* 12(3): e0174157.

Scofield HN, Mattila HR (2015) Honey bee workers that are pollen stressed as larvae become poor foragers and waggle dancers as adults. *PLoSONE* 10(4): e0121731.

Scriven JJ, Whitehorn PR, Goulson D, Tinsley MC (2016) Bergmann's Body Size Rule Operates in Facultatively Endothermic Insects: Evidence from a Complex of Cryptic Bumblebee Species. *PLoS ONE* 11(10): e0163307.

Sekercioglu CH, Riley A (2005) A brief survey of the birds in Kumbira Forest, Gabela, Angola. *Ostrich* 76: 111-117.

Serrano ARM, Capela RA, Santos CV-DN (2017) Biodiversity and notes on carabid beetles from Angola with description of new taxa (Coleoptera:Carabidae). *Zookeys* 4353(2): 201-256.

Shaibi T, Muñoz I, Dall'Olio R, Lodesani M, de la Rúa P, Moritz RFA (2009) *Apis mellifera* evolutionary lineages in Northern Africa: Libya, where orient meets occident. *Insectes Sociaux* 56(3): 293–300.

Sheffield CS, Pindar A, Packer L, Kevan PG (2013) The potential of cleptoparasitic bees as indicator taxa for assessing bee communities. *Apidologie* 44: 501-510.

Shuler R, Roulston TH, Farris G (2005) Farming practices influence wild pollinator populations on squash and pumpkin. *Journal of Economic Entomology* 98: 790–795.

Sikes DS, Copas K, Hirsch T, Longino JT, Schiegel D (2016) On natural history collections, digitized and not: a response to Ferro and Flick. *Zookeys* 618: 145-158.

Simone-Finstrom M, Spivak M (2010) Propolis and bee health: the natural history and significance of resin use by honey bees. *Apidologie* 41: 295-311.

Simone-Finstrom MD, Spivak M (2012) Increased resin collection after parasite challenge: a case of self-medication in honey bees? *PLoS ONE* 7(3): e34601.

Sinervo B, Méndez-de-la-Cruz F, Miles DB, Heulin B, Bastiaans E, Villagrán-Santa Cruz M *et al.* (2010) Erosion of lizard diversity by climate chnage and altered termal niches. *Science* 328: 894-899.

Skelton PH (2019) The freshwater fishes of Angola. In: Huntley BJ, Russo V, Lages F, Ferrand N (eds) Biodiversity of Angola. Science & conservation: a modern synthesis. Springer, Berlin.

Smith RJ, Hines A, Richmond S, Merrick M, Drew A, Fargo R (2000) Altitudinal variation in body size and population density of *Nicrophorus investigator* (Coleoptera: Silphidae). *Environmental Entomology* 29(2): 290-298.

Stabentheiner A, Vollmann J, Kovac H, Crailsheim K (2003) Oxygen consumption and body temperature of active and resting honeybees. *Journal of Insect Physiology* 49: 881–889.

Stabentheiner A, Kovac H, Hetz SK, Kafer H, Stabentheiner G (2012) Assessing honeybee and wasp thermoregulation and energetics – New insights by combination of flow through respirometry with infrared thermography. *Thermochimica Acta* 534: 77–86.

Stabentheiner A, Kovac H (2014) Energetic optimisation of foraging honeybees: flexible change of strategies in response to environmental challenges. *PLoS ONE* 9: e105432.

Steffan-Dewenter I, Potts SG, Packer L, Ghazoul J (2005) Pollinator diversity and crop pollination services are at risk. *Trends in ecology and evolution* 20(12): 651–653.

Stein K, Coulibaly D, Stenchly K, Goetze D, Porembski S, Lindner A, Konaté S, Linsenmair EK (2017) Bee pollination increases yield quantity and quality of cash crops in Burkina Faso, west Africa. *Nature Scientific Reports* 7: 17691.

Stevens VM, Turlure C, Baguette M (2010) A meta-analysis of dispersal in butterflies. *Biological Reviews* 85: 625–642.

Stiassny MLJ, Teugels GG, Hopkins CD (eds) (2007) *The fresh and brackish water fishes of Lower Guinea, West-Central Africa. Collection Faune et Flore Tropicales* 42, Volume 1 and 2. IRD & Muséum National d'Histoire Naturelle, Paris & Musée Royal de l'Afrique Centrale, Tervuren.

Stone GN (1993) Endothermy in the solitary bee *Anthophora plumipes*: independent measures of thermoregulatory ability, costs of warm-up and the role of body size. *Journal of Experimental Biology* 174: 299-320.

Stone GN, Willmer PG (1989) Warm-up rates and body temperatures in bees: the importance of body size, thermal regime and phylogeny. *Journal of Experimental Biology* 147: 303–328.

Stout JC, Nombre I, de Brujin B, Delaney A, Doke DA, Gyimah T, *et al.* (2018) Insect pollination improves yield of Shea (*Vitellaria paradoxa* subsp. *paradoxa*) in the agroforestry parklands of west Africa. *Journal of Pollination Ecology* 22(2): 11-20.

Sunday JM, Bates AE, Dulvy NK (2012) Thermal tolerance and the global redistribution of animals. *Nature Climate Change* 2: 686-690.

Svensson MS, Bersacola E, Mills MSL, Munds RA, Nijman V, Perkin A et al. (2017) A giant among dwarfs: a new species of galago (Primates: Galagidae) from Angola. *American Journal of Physical Anthropology*: 1–14.

Taylor PJ, Neef G, Keith M, Weier S, Monadjem A, Parker DM (2018) Tapping into technology and the biodiversity informatics revolution: updated terrestrial mammal list of Angola with new records from the Okavango basin. *Zookeys* 779: 51-88.

Thomas CD, Cameron A, Green R, Bakkenes M, Beaumont LJ, Collingham YC, *et al.* (2004) Extinction risk from climate change. *Nature* 427 (6970): 145-148.

Thorp RW (2000) The collection of pollen by bees. *Plant systematics and evolution* 222: 211-223.

Torné-Noguera A, Rodrigo A, Arnan X, Osorio S, Barril-Graells H, et al. (2014) Determinants of Spatial Distribution in a Bee Community: Nesting Resources, Flower Resources, and Body Size. *PLoS ONE* 9(5): e97255.

Troia M, McManamay (2017) Completeness and coverage of open-access freshwater fish distribution data in the United States. Diversity and Distributions 23(12).

Tscharntke T, Gathman A, Steffan-Dewenter I (1998) Bioindication using trap-nesting bees and wasps and their natural enemies: community structure and interactions. *Journal of Applied Ecology* 35: 708-719.

Tscharntke T, Klein A, Kruess A, Steffan-Dewenter I, Thies C (2005) Landscape perspectives on agricultural intensification and biodiversity – Ecosystem service management. *Ecology Letters* 8(8): 857-874.

Tscharntke T, Clough Y, Wanger TC, Jackson L, Motzke I, Perfecto I, Vandermeer J, Whitbread A (2012) Global food security, biodiversity conservation and the future of agricultural intensification. *Biological Conservation* 151(1): 53-59.

Tylianakis J, Veddeler D, Lozada T, Lopéz RM, Benítez P, Klein A-M *et al.* (2004) Biodiversity of land-use systems in coastal Ecuador and bioindication using trap-nesting bees, wasps, and their natural enemies. *Lyonia* 6(2): 7-15.

Tylianakis J (2013) The Global Plight of Pollinators. *Science* 339(6127): 1532–1533.

Valtonen A, Molleman F, Chapman CA, Carey JR, Ayres MP, Roininen H (2013) Tropical phenology: bi-annual rhythms and interannual variation in an Afrotropical butterfly assemblage. *Ecosphere* 4(3): 36.

Vanbergen A, IPI (2013) Threats to an ecosystem service: pressures on pollinators. *Frontiers in Ecology and the Environment* 11(5): 251–259.

Vaudo AD, Tooker JF, Grozinger CM, Patch HM (2015) Bee nutrition and floral resource restoration. *Current Opinion in Insect Science* 10: 133-141.

Vazquez DP, Simberloff D (2002) Ecological specialization and susceptibility to disturbance: Conjectures and refutations. *American Naturalist* 159: 606–623.

Veijalainen A, Saaksjarvi IE, Tuomisto H, Broad GR, Bordera S, Jussila R (2014) Altitudinal trends in species richness and diversity of Mesoamerican parasitoid wasps (Hymneoptera: Ichneumonidae). *Insect conservation and diversity* 7: 496-507.

Waser NM, Ollerton J (2006) Plant-pollinator interactions: from specialization to generalization. *The University of Chicago Press*. Chicago.

Weber MG, Mitko L, Eltz T, Ramirez SR (2016) Macroevolution of perfume signalling in orchid bees. *Ecology Letters* 19: 1314-1323.

Werger MJA, Coetzee BJ (1978) Biogeography and ecology in southern Africa. *Junk, The Hague:* 1355

Wcislo WT (1996) Floral resource utilization by solitary bees (Hymenoptera:Apoidea) and exploitation of their stored foods by natural enemies. *Annual review of entomology* 41: 257-286.

Widhiono I, Sudiana E, Darsono D (2017) Diversity of wild bees along elevational gradient in an agricultural area in central Java, Indonesia. *Hindawi Psyche*, article ID 2968414.

Williams NM, Crone EE, Roulston TH, Minckley RL, Packer L, Potts SG (2010) Ecological and life-history trait predict bee species responses to environmental disturbances. *Biological Conservation* 143: 2280-2291.

Winfree R, Aguilar R, Vazquez DP, LeBuhn G, Aizen MA (2009) A meta-analysis of bees' response to anthropogenic disturbance. *Ecology* 90: 2068–2076.

World Wildlife Fund, McGinley M (2008) Angolan scarp savanna and woodlands. In: Cleveland CJ (ed) Encyclopedia of Earth. Environmental Information Coalition, National Council for Science and the Environment, Washington. https://www.worldwildlife.org/ecoregions/at1002. Accessed in December 2018.

Wray JC, Neame LA, Elle E (2014) Floral resources, body size, and surrounding landscape influence bee community assemblages in oak-savannah fragments. *Ecological Entomology* 39: 83-93.

Wright DH (1988) Temporal changes in nectar availability and *Bombus appositus* (Hymenoptera: Apidae) foraging profits. *The Southwestern Naturalist* 33(2): 219-227.

Wu X, Lv M, Jin Z, Michishita R, Chen J, Tian H (2014) Normalized difference vegetation index dynamic and spatiotemporal distribution of migratory birds in the Poyang Lake wetland, China. *Ecology Indicators* 47:217–230.

van Wyk AE, Smith GF (2001) Regions of floristic endemism in southern Africa: a review with emphasis on succulents. *Umdaus Press*, Hatfield.

Zayed A, Packer L (2005) Complementary sex determination substantially increases extinction proneness of haplodiploid populations. *Proceedings of the National Academy of Sciences* 102: 10742-10746.

APPENDIX I - List of bibliography on Angolan beekeeping from the colonial period that were not physically available on the visited Angolan libraries or on digital libraries:

1. Ennis, M.W., Welty, S.F. (1946) Honey hunting in Angola. Nature Mag. 39(4): 185-187.
2. de Portugal Araújo, V. (1954) Existe em Angola um pássaro inimigo das abelhas. A Província de Angola 31(8651): 1-2.
3. de Portugal Araújo, V. (1954) A *Apis ligustica* e a *Apis adansonii*. A Província de Angola 31(8598): 2.
4. de Portugal Araújo, V. (1954) Angola anda a desperdiçar a requeza que as abelhas lhe oferecem. A Província de Angola 31(8629): 1-2.
5. de Portugal Araújo, V. (1956) Notas bionómicas sobre *Apis mellifera adansonii* Latr. Dusenia 7(2): 91-102
6. de Portugal Araújo, V. (1957) Abelhas e mel: a polinização dos cafezais. Gazeta Agrícola de Angola 1(2).
7. de Portugal Araújo, V. (1957) Abelhas e mel: regiões apícolas de Angola. Gazeta Agrícola de Angola 1(3).
8. de Portugal Araújo, V. (1959) Report from Angola. Bee Wld 40: 126-127.
9. de Portugal Araújo, V. (1960) Advances in beekeeping in Angola. Bee Wld 41: 233.
10. de Portugal Araújo, V. (1960) Apiários e técnica apícola africana. Ser. Divulg. Agron. Angolana 18: 16pp.
11. de Portugal Araújo, V. (1960) A colmeia Dadant africana. Ser. Divulg. Agron. Angolana 17: 25pp.
12. Magazine *Abelhas* (1961) Apicultura Angolana. Abelhas 4: 70-71, 76-78.
13. Director Posto Central de Fomento Apícola (1963) Exportação de mel para o ultramar português. Abelhas 37(64): 48.
14. Abreu Lopes, J.A. (1963) Estudo laboratorial da afecção 'loque' europeia na província de Angola. Agro-veterinária 1(2): 4-16.
15. Luis, J.E. (1965) A apicultura e suas vantagens para a agricultura angolana. Gazeta Agrícola de Angola 10(6): 282-284.
16. Magazine *Abelhas* (1971) Colmeias de fibrocimento em Angola. Abelhas 14(158): 23pp.
17. Paixão, V.C. (1971) Aproveitamento dos recursos apícolas de Angola. Abelhas 14(157):5-7; (158):14-16.
18. Magazine *Abelhas* (1973) A indústria apícola pode ser em Angola uma fonte de maior riqueza? Abelhas 16: 12; 17.
19. de Portugal Araújo, V. (1973) Projecto de desenvolvimento apícola para o planalto central 1973/1974 – Bailundo, Angola: Missão de Extensão Rural de Angola iii: 76pp.
20. de Portugal Araújo, V. (1974) Apiários e instalações apícolas na extensão rural (tecnologia apícola) planalto central de Angola – Nova Lisboa, Angola: Estado de Angola ix: 110pp.
21. Missão de Extensão Rural de Angola (1974) Programa de trabalhos 1974. Nova Lisboa, Angola: Missão de Extensão Rural ix: 215pp + 4 maps.
22. Reordenamento (1975) A apicultura, no programa da extensão de Angola. Reordenamento nº 49-50.

Appendix II - Preliminary checklist of bee species (Insecta: Hymenoptera: Antophila) for Angola – bibliographic revision*

* All sp. records were not considered for the final number of bee species. Only one example or online reference is mentioned.
** List of references and online sources used for this revision.

Family Andrenidae

Species list	Reference & Notes	URL
Melitturga capensis Brauns, 1912		https://www.discoverlife.org/mp/20q?search=Melitturga+capensis&flags=subgenus:
Melitturga flavomarginata Patiny, 2000	Eardley & Urban, 2010	http://www.waspweb.org/Apoidea/Andrenidae/Panurginae/Melitturga/index.htm
Meliturgula scriptifrons Walker, 1971	Eardley & Urban, 2010	
	syn. *Meliturgula minima*, Friese	http://www.waspweb.org/Apoidea/Andrenidae/Panurginae/Meliturgula/index.htm

Family Apidae

Species list	Reference & Notes	URL
Allodape derufata Strand	Eardley & Urban, 2010 ? syn. *Allodape mea derufata* Strand, 1912	
Allodape mirabilis Schulz, 1906	? syn. *Allodape mea derufata* Strand, 1912	http://www.gbif.org/occurrence/856590066
Amegilla acraensis Fabricius, 1793	Eardley & Urban, 2010	http://www.waspweb.org/Apoidea/Apidae/Apinae/Anthophorini/Amegilla/index.htm
Amegilla africana Friese, 1905		https://www.gbif.org/occurrence/991930952
Amegilla albocaudata Dours, 1869		https://www.gbif.org/occurrence/991930376
Amegilla atrocincta Lepeletier, 1841	Eardley & Urban, 2010	http://www.waspweb.org/Apoidea/Apidae/Apinae/Anthophorini/Amegilla/index.htm
Amegilla caerulea Friese, 1905	Eardley & Urban, 2010	http://www.waspweb.org/Apoidea/Apidae/Apinae/Anthophorini/Amegilla/index.htm
Amegilla calens Lepeletier, 1841	Eardley & Urban, 2010	http://www.waspweb.org/Apoidea/Apidae/Apinae/Anthophorini/Amegilla/index.htm
Amegilla capensis Friese, 1905	Eardley & Urban, 2010	http://www.waspweb.org/Apoidea/Apidae/Apinae/Anthophorini/Amegilla/index.htm
Amegilla cincta Fabricius, 1781		http://www.gbif.org/occurrence/991930947
Amegilla elsei Brooks, 1988	Eardley & Urban, 2010	http://www.waspweb.org/Apoidea/Apidae/Apinae/Anthophorini/Amegilla/index.htm
Amegilla langi Cockerell, 1935		http://www.gbif.org/occurrence/1275016633
Amegilla mimadvena Cockerell, 1916	Results from this study	
Amegilla nubica Lepeletier, 1841	syn. *Anthophora nubica* Lepeletier, 1841	https://www.gbif.org/occurrence/1275016552
Amegilla obscuriceps Friese, 1905		http://www.gbif.org/occurrence/1275016553
Amegilla paradoxa Brooks, 1988		http://www.gbif.org/occurrence/991930358
Amegilla penicula Eardley, 1994	Results from this study	

Family Apidae Species list	Reference & Notes	URL
Amegilla spilostoma Cameron, 1905		http://www.gbif.org/occurrence/1275016574
Amegilla velutina Friese, 1909		http://www.gbif.org/occurrence/1275016631
Amegilla vivida Smith, 1879		http://www.gbif.org/occurrence/991930986
Anthophora angolensis Dalla Torre, 1896	Eardley & Urban, 2010	https://www.discoverlife.org/mp/20l?id=AMNH_BEES121804
Anthophora armata Friese, 1905	Eardley & Urban, 2010	https://www.discoverlife.org/mp/20l?id=AMNH_BEES121837
Anthophora auone Eardley and Brooks, 1989	Eardley & Urban, 2010	https://www.gbif.org/occurrence/1275010600
Anthophora joetta Brooks, 1988	Eardley & Urban, 2010	https://www.discoverlife.org/mp/20l?id=AMNH_BEES122752
Anthophora oldi Meade-Waldo, 1914	Eardley & Urban, 2010	https://www.gbif.org/occurrence/1092726361
Anthophora rufolanata Dours, 1869		http://www.gbif.org/occurrence/1275016577
Anthophora vestita Smith, 1854	Eardley & Urban, 2010	https://www.gbif.org/occurrence/1275016583
Apis mellifera adansonii Latreille, 1804	Rosário-Nunes & Tordo, 1960	
Apis mellifera unicolor Latreille, 1802	Rosário-Nunes & Tordo, 1960	
Apis mellifera intermissa Linnaeus, 1758	Rosário-Nunes & Tordo, 1960	
Braunsapis albipennis Friese, 1909		http://www.gbif.org/occurrence/1275016643
Braunsapis albitarsis Friese, 1924	syn. *Paratetrapedia albitarsis*	http://www.gbif.org/occurrence/856592509
Braunsapis angolensis Cockerell,1933	Eardley & Urban, 2010	http://www.waspweb.org/Apoidea/Apidae/Xylocopinae/Allodapini/Braunsapis/index.htm
Braunsapis bouyssoui Vachal, 1903		http://www.waspweb.org/Apoidea/Apidae/Xylocopinae/Allodapini/Braunsapis/index.htm
Braunsapis calidula Cockerell, 1908	Eardley & Urban, 2010	http://www.waspweb.org/Apoidea/Apidae/Xylocopinae/Allodapini/Braunsapis/index.htm
Braunsapis facialis Gerstaecker, 1858	Eardley & Urban, 2010	http://www.waspweb.org/Apoidea/Apidae/Xylocopinae/Allodapini/Braunsapis/index.htm
Braunsapis foveata Smith, 1854	Eardley & Urban, 2010	http://www.waspweb.org/Apoidea/Apidae/Xylocopinae/Allodapini/Braunsapis/index.htm
Braunsapis langenburgensis Strand, 1915		http://www.gbif.org/occurrence/287263017
Braunsapis leptozonia Vachal, 1909	Eardley & Urban, 2010	http://www.waspweb.org/Apoidea/Apidae/Xylocopinae/Allodapini/Braunsapis/index.htm
Braunsapis natalica Michener, 1970	Eardley & Urban, 2010	http://www.waspweb.org/Apoidea/Apidae/Xylocopinae/Allodapini/Braunsapis/index.htm
Braunsapis otavica Cockerell, 1939		https://www.gbif.org/occurrence/1275022456
Braunsapis rhodesi Cockerell, 1936		http://www.gbif.org/occurrence/287263103
Braunsapis rolini Vachal, 1903	Results from this study	
Braunsapis vitrea Vachal, 1903	Eardley & Urban, 2010	http://www.waspweb.org/Apoidea/Apidae/Xylocopinae/Allodapini/Braunsapis/index.htm
Ceratina aereola Vachal, 1903	Eardley & Urban, 2010	https://www.discoverlife.org/mp/20l?id=AMNH_BEES107125

Family Apidae
Species list

Species list	Reference & Notes	URL
Ceratina aloes Cockerell, 1932	Results from this study	
Ceratina ericia Vachal, 1903	Eardley & Urban, 2010	https://www.gbif.org/occurrence/1275016581
Ceratina inermis Friese, 1905	Results from this study	
Ceratina labrosa Friese, 1905		http://www.gbif.org/occurrence/1275016601
Ceratina linola Vachal, 1903	Results from this study	
Ceratina moerenhouti Vachal, 1903	Eardley & Urban, 2010	https://www.discoverlife.org/mp/20l?id=AMNH_BEES108346
Ceratina nasalis Friese, 1905	Results from this study	
Ceratina nigriceps Friese, 1905	Results from this study	
Ceratina nyassensis Strand, 1912	Results from this study	
Ceratina pennicillata Friese, 1905	Eardley & Urban, 2010	
Ceratina rossi Daly, 1988	Eardley & Urban, 2010	https://www.discoverlife.org/mp/20l?id=AMNH_BEES108677
Ceratina rufogastra Cockerell, 1937	Results from this study	
Ceratina tanganycensis Strand, 1911	Results from this study	
Ceratina viridis Guèrin, 1845	Eardley & Urban, 2010	
Cleptotrigona cubiceps Friese, 1912	Eardley & Urban, 2010 Syn. *Lestrimellita cubiceps*	http://www.gbif.org/occurrence/1092734210
Compsomelissa stigmoides Michener, 1971	Results from this study	
Dactylurina staudingeri Gribodo, 1893	Eardley & Urban, 2010	https://www.gbif.org/occurrence/1092734354
Hypotrigona araujoi Michener, 1959	Eardley & Urban, 2010	http://www.waspweb.org/Apoidea/Apidae/Apinae/Meliponini/Hypotrigona/index.htm
Hypotrigona gribodoi Magretti, 1884	Eardley & Urban, 2010	http://www.waspweb.org/Apoidea/Apidae/Apinae/Meliponini/Hypotrigona/index.htm
Hypotrigona ruspolii Magretti, 1898	Eardley & Urban, 2010 Syn. *Liotrigona bouyssoui*	http://www.waspweb.org/Apoidea/Apidae/Apinae/Meliponini/Hypotrigona/index.htm
Liotrigona bottegoi Magretti, 1895	Eardley & Urban, 2010	http://www.gbif.org/occurrence/1092738520
Liotrigona parvula Darchen, 1971		https://www.gbif.org/occurrence/1092739696
Macrogalea candida Smith, 1879	Eardley & Urban, 2010	https://www.gbif.org/occurrence/1275016570
Meliponula beccarii Gribodo, 1879	Eardley & Urban, 2010 Syn. *Meliplebeia beccarii*	http://www.waspweb.org/Apoidea/Apidae/Apinae/Meliponini/Meliponula/index.htm
Meliponula bocandei Spinola, 1853	Eardley & Urban, 2010 syn. *Melipona bocandei* Spinola, 1853	http://www.waspweb.org/Apoidea/Apidae/Apinae/Meliponini/Meliponula/index.htm
Meliponula ferruginea Lepeletier, 1841	Eardley & Urban, 2010 de Portugal-Araújo, 1955	http://www.waspweb.org/Apoidea/Apidae/Apinae/Meliponini/Meliponula/index.htm

Family Apidae Species list	Reference & Notes	URL
	syn. *Axestotrigona ferruginea; Meliponula erythra*	
Meliponula lendliana Friese, 1900	Eardley & Urban, 2010	http://www.waspweb.org/Apoidea/Apidae/Apinae/Meliponini/Meliponula/index.htm
Pasites dichroa Smith, 1854	Eardley & Urban, 2010	https://www.gbif.org/occurrence/658522953
Tetralonia caudata Friese, 1905	Results from this study	
Tetralonia cinctula Cockerell, 1936		http://www.gbif.org/occurrence/1275016558
Tetralonia macrognatha Gerstäcker, 1870	Eardley & Urban, 2010	https://www.gbif.org/occurrence/1275016588
Tetralonia penicillata Friese, 1905		http://www.gbif.org/occurrence/1092787498
Tetraloniella watmoughi Eardley, 1989	Eardley & Urban, 2010	https://www.discoverlife.org/mp/20l?id=AMNH_BEES120235
Thyreus abyssinicus Radoszkowski, 1873	Eardley & Urban, 2010	https://www.gbif.org/occurrence/894574802
Thyreus bouyssoui Vachal, 1903	Eardley & Urban, 2010	https://www.discoverlife.org/mp/20l?id=AMNH_BEES125047
Thyreus calceatus Vachal, 1903		https://www.gbif.org/occurrence/1275016609
Thyreus meripes Vachal, 1903	Eardley & Urban, 2010	https://www.gbif.org/occurrence/894574834
Thyreus niloticus Cockerell, 1937		https://www.gbif.org/occurrence/894574913
Thyreus pictus Smith, 1854	Eardley & Urban, 2010	https://www.gbif.org/occurrence/894575022
Thyreus somalicus **Strand, 1911**	Results from this study	
Thyreus tschoffeni Vachal, 1903	Eardley & Urban, 2010	https://www.gbif.org/occurrence/894575132
Thyreus vachali **Friese, 1905**	Results from this study	
Xylocopa africana Fabricius, 1781	Eardley & Urban, 2010	https://www.discoverlife.org/mp/20q?search=Xylocopa+africana&flags=subgenus:
Xylocopa albiceps Fabricius, 1804	Eardley & Urban, 2010	https://www.discoverlife.org/mp/20q?act=x_ant&path=Insecta/Hymenoptera/Apoidea/Apidae/Xylocopa/albiceps&name=Xylocopa+albiceps&guide=Bees_Ghana&authority=Fabricius,+1804
Xylocopa caffra Linnaeus, 1767	Eardley & Urban, 2010	https://www.gbif.org/pt/occurrence/2029380639
Xylocopa calens Lepeletier, 1841	Eardley & Urban, 2010	https://www.discoverlife.org/mp/20l?id=GBIF287272661
Xylocopa chiyakensis Cockerell, 1908	Eardley & Urban, 2010	
Xylocopa combusta Smith, 1854	Eardley & Urban, 2010	https://www.gbif.org/occurrence/2029385325
Xylocopa erythrina Gribodo, 1894		https://www.discoverlife.org/mp/20l?id=GBIF287272904
Xylocopa flavicollis DeGeer, 1778	syn. *Xylocopa divisa*	https://www.gbif.org/pt/occurrence/2443208847
Xylocopa flavorufa DeGeer, 1778	Eardley & Urban, 2010	https://www.gbif.org/occurrence/2029380855
Xylocopa hottentotta Smith, 1854	Eardley & Urban, 2010	

Family Apidae

Species list	Reference & Notes	URL
Xylocopa imitator Smith, 1854	Eardley & Urban, 2010	https://www.discoverlife.org/mp/20l?id=GBIF287273620
Xylocopa inconstans Smith, 1874	Eardley & Urban, 2010	https://www.discoverlife.org/mp/20l?id=AMNH_BEES105908
Xylocopa lugubris Gerstacker, 1857	Eardley & Urban, 2010	https://www.gbif.org/occurrence/2443798735
Xylocopa mixta Radoszkowski, 1881	Eardley & Urban, 2010	https://www.gbif.org/occurrence/894565349
Xylocopa modesta Smith, 1854	Eardley & Urban, 2010	https://www.discoverlife.org/mp/20l?id=AMNH_BEES106206
Xylocopa nigrita Fabricius, 1775	Eardley & Urban, 2010	https://www.discoverlife.org/mp/20l?id=AMNH_BEES106291
Xylocopa obscurata Smith, 1854	Eardley & Urban, 2010	https://www.discoverlife.org/mp/20l?id=AMNH_BEES106313
Xylocopa olivacea Fabricius, 1778	Eardley & Urban, 2010	https://www.gbif.org/occurrence/2029379289
Xylocopa orthosiphonis Cockerell, 1908	Eardley & Urban, 2010	https://www.discoverlife.org/mp/20l?id=AMNH_BEES106368
Xylocopa rufitarsis Lepeletier, 1841	Eardley & Urban, 2010	https://www.discoverlife.org/mp/20l?id=AMNH_BEES106543
Xylocopa senior Vachal, 1899	Eardley & Urban, 2010	https://www.discoverlife.org/mp/20l?id=AMNH_BEES106593
Xylocopa torrida Westwood, 1838	Eardley & Urban, 2010	https://www.discoverlife.org/mp/20l?id=AMNH_BEES106706
Xylocopa varipes Smith, 1854		https://www.gbif.org/occurrence/894564674
Xylocopa wellmani Cockerell, 1906	Eardley & Urban, 2010	https://www.discoverlife.org/mp/20l?id=AMNH_BEES107060

Family Colletidae

Species List	Reference & Notes	URL
Colletes spp		https://www.gbif.org/occurrence/2029380381
Hylaeus binotatus Alfken, 1914	Eardley & Urban, 2010	http://www.waspweb.org/Apoidea/Colletidae/Hylaeinae/Hylaeus/index.htm
Hylaeus cribratus Bridwell, 1919	Eardley & Urban, 2010	http://www.waspweb.org/Apoidea/Colletidae/Hylaeinae/Hylaeus/index.htm

Family Halictidae

Species list	Reference & Notes	URL
Cellariella kalaharica Cockerell, 1936	Eardley & Urban, 2010	https://www.discoverlife.org/mp/20q?search=Cellariella+kalaharica&flags=subgenus:
Cellariella somalica Magretti, 1899	Eardley & Urban, 2010	https://www.discoverlife.org/mp/20q?search=Cellariella+somalica&flags=subgenus:&mobile=1
Ceylalictus congoensis Pesenko & Pauly, 2005	Eardley & Urban, 2010	http://www.waspweb.org/Apoidea/Halictidae/Nomioidinae/Ceylalictus/index.htm
Ceylalictus muiri Cockerell, 1909	Eardley & Urban, 2010	https://www.discoverlife.org/mp/20q?search=Ceylalictus+muiri&flags=col2:subgenus:
Eupetersia spp	Results from this study	

Family Halictidae

Species list	Reference & Notes	URL
Halictus geigeriae Cockerell, 19808	Eardley & Urban, 2010	
Lasioglossum creightoni Cockerell, 1908	Eardley & Urban, 2010	http://www.waspweb.org/Apoidea/Halictidae/Halictinae/Lasioglossum/index.htm
Lasioglossum michaelseni Friese, 1916	Eardley & Urban, 2010	http://www.waspweb.org/Apoidea/Halictidae/Halictinae/Lasioglossum/index.htm
Lipotriches cirrita Vachal, 1903	Eardley & Urban, 2010	https://www.discoverlife.org/mp/20q?search=Lipotriches+cirrita&flags=subgenus:
Lipotriches collaris Vachal, 1903	Pauly, 2014	https://www.discoverlife.org/mp/20q?search=Lipotriches+collaris&flags=subgenus:
Lipotriches hylaeoides Gerstäcker, 1857	Pauly, 2014; Eardley & Urban, 2010	https://www.discoverlife.org/mp/20q?search=Lipotriches+hylaeoides&flags=subgenus:
Lipotriches natalensis Cockerell, 1916	Eardley & Urban, 2010 syn. *Lipotriches johannis* Cockerell, 1916	https://www.discoverlife.org/mp/20q?act=x_ant&path=Insecta/Hymenoptera/Apoidea/Halictidae/Lipotriches/natalensis&name=Lipotriches+natalensis&guide=Bees_Ghana&authority=(Cockerell,+1916)&mobile=close
Lipotriches notabilis notabilis Schletterer, 1891	Pauly, 2014	https://www.gbif.org/species/100250145
Lipotriches panganina **Pauly, 1990**		https://www.gbif.org/occurrence/1224580761
Lipotriches predonta Pauly, 2014	Pauly, 2014	https://www.discoverlife.org/mp/20q?search=Lipotriches+predonta&flags=subgenus:
Lipotriches tridentata Smith, 1875	Eardley & Urban, 2010	https://www.discoverlife.org/mp/20q?search=Lipotriches+tridentata&flags=subgenus:
Lipotriches welwitschi Cockerell, 1908	Pauly, 2014; Eardley & Urban, 2010	https://www.discoverlife.org/mp/20q?search=Lipotriches+welwitschi&flags=col5:subgenus:
Nomia amabilis Cockerell, 1908	Eardley & Urban, 2010	https://www.discoverlife.org/mp/20q?search=Nomia+amabilis&flags=subgenus:
Nomia rufitarsis Smith, 1875	Eardley & Urban, 2010	https://www.discoverlife.org/mp/20q?search=Nomia+rufitarsis&flags=subgenus:
Nomia scitula Bingham, 1903	Eardley & Urban, 2010	https://www.discoverlife.org/mp/20q?search=Nomia+scitula&flags=subgenus:
Pseudapis aliceae Cockerell, 1935	Eardley & Urban, 2010	https://www.discoverlife.org/mp/20q?search=Pseudapis+aliceae&flags=col5:subgenus:&mobile=1
Pseudapis anthidioides Gerstäcker, 1857	Eardley & Urban, 2010	https://www.discoverlife.org/mp/20q?search=Pseudapis+anthidioides&flags=subgenus:
Pseudapis cinerea Friese, 1930	Eardley & Urban, 2010	https://www.discoverlife.org/mp/20q?search=Pseudapis+cinerea&flags=subgenus:
Pseudapis interstitinervis Strand, 1912	Eardley & Urban, 2010	https://www.discoverlife.org/mp/20q?search=Pseudapis+interstitinervis&flags=subgenus:
Pseudapis usakoa Cockerell, 1939	Eardley & Urban, 2010	https://www.discoverlife.org/mp/20q?search=Pseudapis+usakoa&flags=col4:subgenus:
Seladonia centrosa Vachal, 1903	Eardley & Urban, 2010	
Seladonia hotoni Vachal, 1903	Eardley & Urban, 2010	http://www.atlashymenoptera.net/pagetaxon.asp?tx_id=1764
Seladonia jucundus Smith, 1853	Eardley & Urban, 2010 syn. *Halictus jucundus*	https://www.discoverlife.org/mp/20q?search=Halictus+jucundus&flags=subgenus:
Seladonia seminiger Cockerell, 1937	Eardley & Urban, 2010 syn. *Halictus seminiger*	https://www.discoverlife.org/mp/20q?search=Halictus+seminiger&flags=subgenus:
Sphecodes spp Latreille, 1804		http://www.gbif.org/occurrence/785001778
Thrinchostoma orchidarum Cockerell, 1908	Eardley & Urban, 2010	http://www.waspweb.org/Apoidea/Halictidae/Halictinae/Thrinchostoma/index.htm
Thrinchostoma othonnae Cockerell, 1908	Eardley & Urban, 2010	http://www.waspweb.org/Apoidea/Halictidae/Halictinae/Thrinchostoma/index.htm

Family Halictidae

Species list	Reference & Notes	URL
Thrinchostoma wellmanni Cockerell, 1908	Eardley & Urban, 2010	http://www.waspweb.org/Apoidea/Halictidae/Halictinae/Thrinchostoma/index.htm

Family Megachilidae

Species list	Reference & Notes	URL
Afranthidium minutulum Friese, 1905	Eardley & Urban, 2010	
Afroheriades spp	Results from this study	
Anthidiellum eritrinum Friese, 1915	Results from this study	
Anthidiellum polyochrum Mavromoustakis, 1937		http://www.gbif.org/occurrence/1275016627
Anthidioma spp	Results from this study	
Anthidium severini Vachal, 1903	syn. *Anthidium severini eriksoni* Mavromoustakis	http://www.gbif.org/occurrence/1224583750
Coelioxys planidens Friese, 1904		https://www.discoverlife.org/mp/20l?id=AMNH_BEES99084
Coelioxys setosa Friese, 1904	Eardley & Urban, 2010	https://www.discoverlife.org/mp/20q?search=Coelioxys+setosus&guide=Coelioxys_female&flags=subgenus:
Coelioxys torrida Smith, 1854	Eardley & Urban, 2010	https://www.discoverlife.org/mp/20q?act=x_ant&path=Insecta/Hymenoptera/Apoidea/Megachilidae/Coelioxys/torrida&name=Coelioxys+torrida&authority=Smith,+1854
Euaspis abdominalis Fabricius, 1793	Eardley & Urban, 2010	https://www.discoverlife.org/mp/20q?search=Euaspis+abdominalis&flags=subgenus:&mobile=1
Heriades communis Cockerell, 1931	Eardley & Urban, 2010	
Heriades wellmani Cockerell, 1908	Eardley & Urban, 2010	https://www.discoverlife.org/mp/20q?search=Heriades+wellmani&mobile=close&flags=glean:
Megachile admixta Cockerell, 1931	Eardley & Urban, 2010	https://www.discoverlife.org/mp/20q?search=Megachile+admixta&flags=subgenus:&mobile=1
Megachile angolensis Cockerell, 1935	Eardley & Urban, 2010	http://www.waspweb.org/Apoidea/Megachilidae/Megachilinae/Megachilini/Megachile/index.htm
Megachile aurifacies Pasteels, 1985	Eardley & Urban, 2010	https://www.discoverlife.org/mp/20q?search=Megachile+aurifacies&flags=col2:subgenus:&mobile=1
Megachile aurifera Cockerell, 1935	Eardley & Urban, 2010	https://www.discoverlife.org/mp/20q?search=Megachile+aurifera&flags=col2:subgenus:
Megachile basalis Smith, 1853		http://www.gbif.org/occurrence/1224584503
Megachile benitocola Strand, 1912		http://www.gbif.org/occurrence/856611952
Megachile biloba Vachal, 1910	Eardley & Urban, 2010	
Megachile bituberculata Ritsema, 1880	Eardley & Urban, 2010	http://www.gbif.org/occurrence/856608379
Megachile bituberculata rubripedana Strand, 1914	Eardley & Urban, 2010	

Family Megachilidae Species list	Reference & Notes	URL
Megachile bombiformis Gerstäcker, 1857	Eardley & Urban, 2010	https://www.discoverlife.org/mp/20q?search=Megachile+bombiformis&flags=subgenus:
Megachile bucephala Fabricius, 1793	Results from this study	
Megachile caricina Cockerell, 1907	Eardley & Urban, 2010	https://www.discoverlife.org/mp/20q?search=Megachile+caricina&guide=Megachilidae_genera&flags=subgenus:&mobile=1
Megachile cincta Fabricius, 1781	Eardley & Urban, 2010	
Megachile cognata Smith, 1853	Eardley & Urban, 2010	https://www.discoverlife.org/mp/20q?search=Megachile+cognata&flags=col2:glean:subgenus:
Megachile curtula Gerstäcker, 1857	syn. *Megachile sticeae*	http://www.gbif.org/occurrence/131340466
Megachile decemsignata Radoszkowski, 1881	Eardley & Urban, 2010	https://www.discoverlife.org/mp/20q?search=Megachile+decemsignata&flags=subgenus:
Megachile devexa Vachal, 1903	syn. Chalicodoma duponti	http://www.gbif.org/occurrence/856600608
Megachile ekuivella Cockerell, 1909	Eardley & Urban, 2010 syn. *Megachile frontalis*	
Megachile eurimera Smith, 1854		http://www.gbif.org/occurrence/856609855
Megachile felina Gerstäcker, 1857	Eardley & Urban, 2010	
Megachile fimbriata Smith, 1853		http://www.gbif.org/occurrence/1224585166
Megachile flavipollex **Pasteels, 1960**		https://www.gbif.org/occurrence/1224585274
Megachile fuscicauda Cockerell, 1933	Eardley & Urban, 2010	https://www.discoverlife.org/mp/20q?search=Megachile+fuscicauda&flags=glean:subgenus:&mobile=1
Megachile gowdeyi Cockerell, 1931		http://www.gbif.org/occurrence/856609913
Megachile gratiosa Gerstäcker, 1857	Eardley & Urban, 2010	https://www.discoverlife.org/mp/20q?search=Megachile+gratiosa&guide=Bee_genera&flags=subgenus:&mobile=1
Megachile harthulura Cockerell, 1937		http://www.gbif.org/occurrence/856609993
Megachile hypopyrrha Cockerell, 1937		http://www.gbif.org/occurrence/856610002
Megachile ikuthaensis Friese, 1903	Eardley & Urban, 2010	https://www.discoverlife.org/mp/20q?search=Megachile+ikuthaensis&guide=Apoidea_species&cl=GH&flags=HAS:
Megachile insolita Pasteels, 1965	syn. *Chalicodoma (Chalicodoma) insolita*	https://www.gbif.org/pt/occurrence/1224584540
Megachile khamana Cockerell, 1938	Results from this study	
Megachile malangensis Friese, 1904	Eardley & Urban, 2010	https://www.discoverlife.org/mp/20q?search=Megachile+malangensis&guide=Bee_genera_United_States_and_Canada&flags=subgenus:
Megachile marshalli Friese, 1904	Eardley & Urban, 2010	https://www.discoverlife.org/mp/20q?search=Megachile+marshalli&flags=subgenus:
Megachile maxillosa **Guèrin-Méneville, 1845**		https://www.gbif.org/occurrence/1224584612
Megachile nasicornis Friese, 1903	Results from this study	
Megachile nigripollex Vachal, 1910		http://www.gbif.org/occurrence/856601675
Megachile pallida Radoszkowski, 1881	Eardley & Urban, 2010	https://www.discoverlife.org/mp/20q?search=Megachile+pallida&flags=col3:subgenus:

Family Megachilidae Species list	Reference & Notes	URL
Megachile patellimana Spinola, 1838		http://www.gbif.org/occurrence/856610212
Megachile praetexta Vachal, 1910		http://www.gbif.org/occurrence/856605778
Megachile pulchrifrons Cockerell, 1933	Eardley & Urban, 2010	https://www.discoverlife.org/mp/20q?search=Megachile+pulchrifrons&flags=col3:subgenus:
Megachile pulvinata Vachal, 1910	Eardley & Urban, 2010	https://www.discoverlife.org/mp/20q?search=Megachile+pulvinata&flags=subgenus:
Megachile regina Cheesman, 1938		http://www.gbif.org/occurrence/1224585541
Megachile rufa Friese, 1903	syn. *Creightonella rufa* Friese	http://www.gbif.org/occurrence/856607978
Megachile rufigaster Cockerell, 1945		http://www.gbif.org/occurrence/856612006
Megachile rufipennis **Fabricius, 1804**		https://www.gbif.org/occurrence/1224584929
Megachile semierma Vachal, 1903		http://www.gbif.org/occurrence/856610422
Megachile semivenusta Cockerell, 1931		http://www.gbif.org/occurrence/856612023
Megachile sinuata Friese, 1903	Eardley & Urban, 2010	https://www.discoverlife.org/mp/20q?act=x_ant&path=Insecta/Hymenoptera/Apoidea/Megachilidae/Megachile/sinuata&name=Megachile+sinuata&guide=Bees_Ghana&authority=Friese,+1903&mobile=close
Megachile striatula Cockerell, 1931	Eardley & Urban, 2010	http://www.waspweb.org/Apoidea/Megachilidae/Megachilinae/Megachilini/Megachile/index.htm
Megachile unifasciata Radoszkowski, 1881	Eardley & Urban, 2010	http://www.waspweb.org/Apoidea/Megachilidae/Megachilinae/Megachilini/Megachile/index.htm
Megachile ungulata Smith, 1853		https://www.discoverlife.org/mp/20q?search=Megachile+ungulata&flags=subgenus:
Noteriades ekuivensis Cockerell, 1908	Eardley & Urban, 2010	
Ocheriades spp	Results from this study	
Othinosmia spp	Results from this study	
Pachyanthidium benguelensis Vachal, 1903	Eardley & Urban, 2010	
Pachyanthidium bipectinatum	Results from this study	
Pachyanthidium bouyssoui Vachal, 1903	Eardley & Urban, 2010	https://www.discoverlife.org/mp/20q?search=Pachyanthidium+bouyssoui&flags=subgenus:
Pachyanthidium nigrum Pasteels, 1984	Results from this study	
Pachyanthidium otavicum	Results from this study	
Pachyanthidium spilognathum	Results from this study	
Pseudoanthidium truncatum Smith, 1854	Eardley & Urban, 2010	https://www.discoverlife.org/mp/20q?act=x_ant&path=Insecta/Hymenoptera/Apoidea/Megachilidae/Pseudoanthidium/truncatum&name=Pseudoanthidium+truncatum&guide=Bees_Ghana&authority=(Smith,+1854)&mobile=close
Serapista denticulata Smith, 1854		http://www.gbif.org/occurrence/1275016658
Trachusa schoutedeni Vachal, 1910	Eardley & Urban, 2010	https://www.discoverlife.org/mp/20q?search=Trachusa+schoutedeni&guide=Bee_genera&flags=subgenus:&mobile=close

Family Megachilidae

Species list	Reference & Notes	URL
Wainia spp	Results from this study	

****Online Sources**

1 Discover Life Bee Species Guide - http://www.discoverlife.org/mp/20q?guide=Apoidea_species
2 South African National Biodiversity Institute: ARC Bees
3 WaspWeb - www.waspweb.org
4 University of Kansas Biodiversity Institute: Snow Entomological Museum Collection
5 Collection of CAS – California Academy of Sciences - http://research.calacademy.org/ent
6 Atlas Hymenoptera - http://www.atlashymenoptera.net/default.asp
7 Naturalis Biodiversity Center: Naturalis Biodiversity Center (NL) - Hymenoptera
8 South African National Biodiversity Institute: Iziko
9 Collection of Smithsonian National Museum of Natural History (NMNH) - http://collections.nmnh.si.edu/search/ento/
10 South African National Biodiversity Institute: SABCA
11 Centro de Referência em Informação Ambiental: Coleção Entomológica Paulo Nogueira-Neto - IB/USP
12 Bee diversity in Angola and community change along na altitudinal gradient at Serra da Chela (Bruco) – results from this study. Elizalde S, van Noort S, Picker MD (2019)

****List of literature used as reference on the online databases and in this revision**

1 Brothers D.J. 1999. Phylogeny and evolution of wasps, ants and bees (Hymenoptera, Chrysidoidea, Vespoidea and Apoidea) Zoologica Scripta 28: 233–250.
2 Eardley, C, Finnamore, A.T. & Michener, C.D. 1993. Superfamily Apoidea (pp. 279-357). In GOULET, H. & HUBER, J. (eds). Hymenoptera of the World: an identification guide to families. Research Branch, Agriculture Canada, Ottawa, Canada, 668 pp.
3 Eardley, C.D. 2004. Taxonomic revision of the African stingless bees (Apoidea: Apidae: Apinae: Meliponini). African Plant Protection 10: 63–96.
4 Eardley, C & Urban, R. 2010. Catalogue of Afrotropical bees (Hymenoptera: Apoidea: Anthophila). Zootaxa 2455: 1-548.
5 Michener, C.D. 2000. The Bees of the World. Johns Hopkins University Press. 953 pp.
6 Pauly, A. 2014. Les Abeilles des Graminées ou Lipotriches Gerstaecker, 1858, sensu stricto (Hymenoptera : Apoidea : Halictidae : Nomiinae) de l'Afrique subsaharienne. Belgian Journal of Entomology, 20 : 1-393.
7 Pauly, A. 2008. Catalogue of the sub-Saharan species of the genus Seladonia Robertson, 1918, with description of two new species (Hymenoptera: Apoidea: Halictidae). Zool. Med. Leiden 82: 391-400.
8 Eardley C. 2013. A taxonomic revision of the southern African leaf-cutter bees, *Megachile* Latreille *sensu stricto* and *Heriadopsis* Cockerell (Hymenoptera: Apoidea: Megachilidae). *Zootaxa* 3601 (1): 001–133

Appendix III - Correlation of environmental variables (annual mean temperature and annual precipitation) and landscape variables at two dimensional scales (mean NDVI and mean Tree cover, 200 and 400 m radii) with altitude.

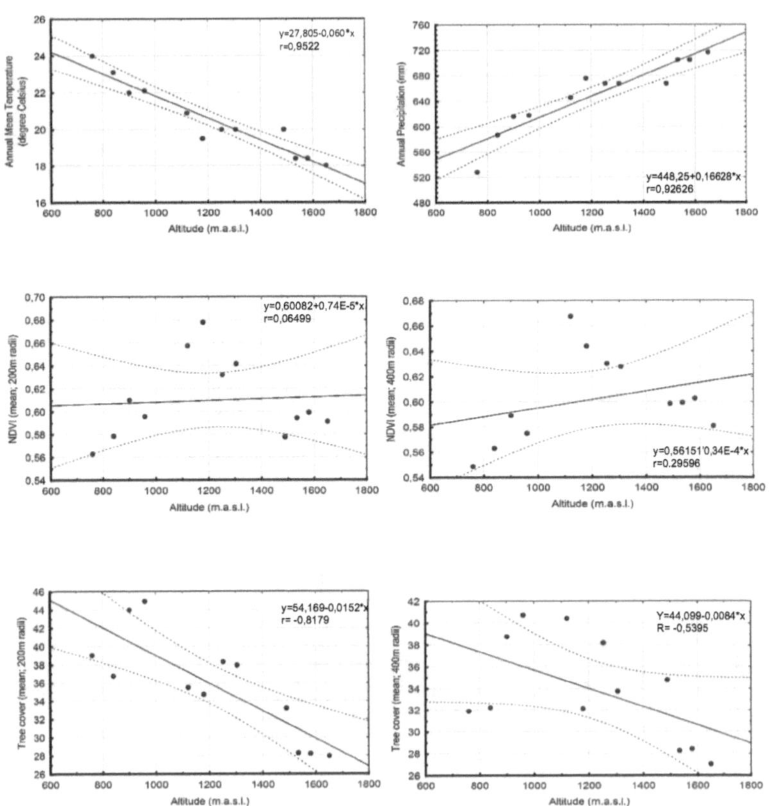

Appendix IV – Bee generic occurrences and distribution along an altitudinal gradient in Bruco – Serra da Chela, Angola.

Table 3.1 – Individuals per genus and site along the altitudinal gradient, including all sampling methods with exception of malaise trap.

Genera	Total specimens collected	Collected specimens per altitude											
		760	840	901	960	1120	1179	1253	1305	1489	1534	1580	1651
Afranthidium	4						1	1			2		
Afroheriades	5	1			1			2					1
Amegilla	5			1			1						3
Anthidiellum	24	4		3	3	1	3	1					
Anthidioma	3						1	1			1		
Apis	16					6	4		1	3		2	
Braunsapis	82	19	11	2	3	1	4	21	4	2	1	5	
Ceratina	98	28	6	4	2	2	15	7	6		21	3	4
Cleptotrigona	2								1	1			
Coelioxys	4				1			2			1		
Colletes	3					2						1	
Eupetersia	5	4			1								
Halictus	15	2				3	2	1			7		
Heriades	41		2		1	4	12	12		1	4	5	
Hypotrigona	191	6	5	45	27	1	6	33	14	7		1	1
Lasioglossum	4	1				1		1			1		
Lipotriches	56	2	2	1		15	1	3	2	4	4	8	5
Macrogalea	93	7	5	7	1	7	4	9	5	3	25	17	3
Megachile	18	1	1	1				1	3		5	5	1
Meliponula	25	2				4	1	6	1	11			
Meliturgula	1					1							
Nomia	10	1	2			1		2				1	3
Noteriades	3	1	2										
Ocheriades	1		1										
Othinosmia	1								1				
Pachyanthidium	51	1	22	1	1	4	2	4	2	3	1	1	
Patellapis	38	3	3	3	1	3	2	1			16	2	5
Pseudoanthidium	12	1				2	1	5			1	2	
Seladonia	10						1	1			5	1	2
Tetralonia	10			7			2					1	
Thrinchostoma	1					1							
Thyreus	6							3			1	1	1
Wainia	2										1	1	
Xylocopa	4			1								1	2
Total	845	84	107	84	43	77	72	117	40	37	97	57	30

Appendix V - Species accumulation curve for bee genera, per sampling site and for the entire transect. Analysis using Jackknife 1 estimates.

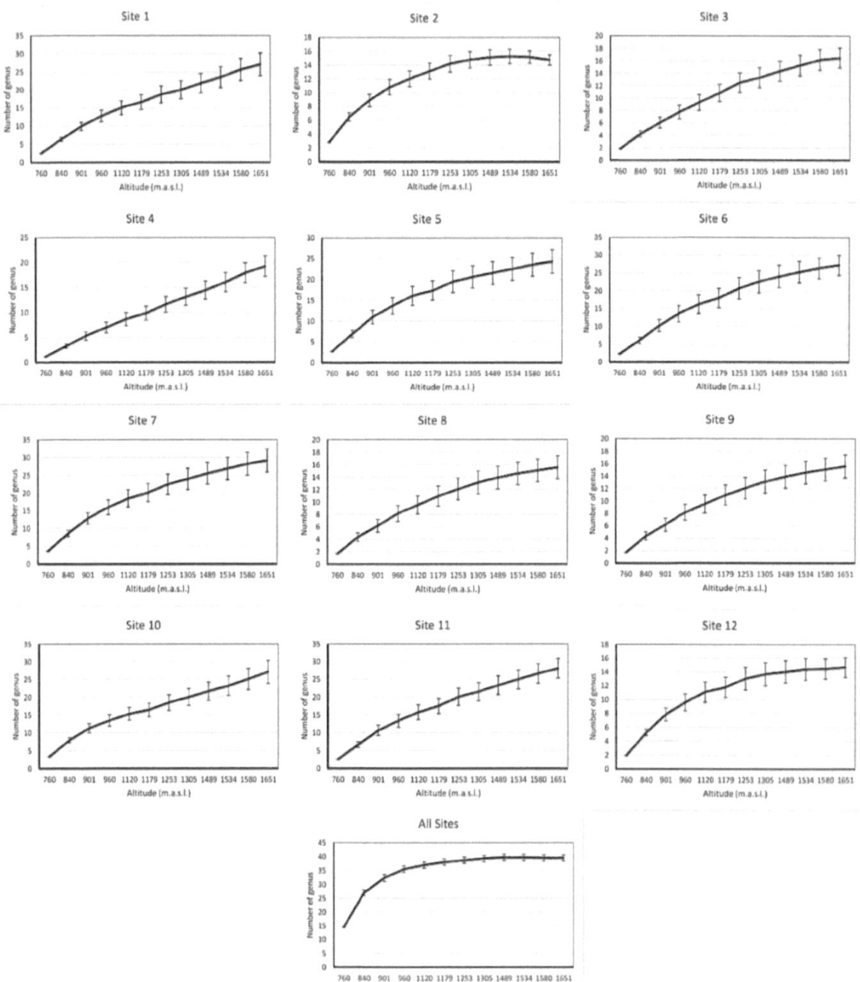

Appendix VI - Results of correlations between diversity measures and altitude (m.a.s.l.), annual precipitation (mm) and annual mean temperature (°C).

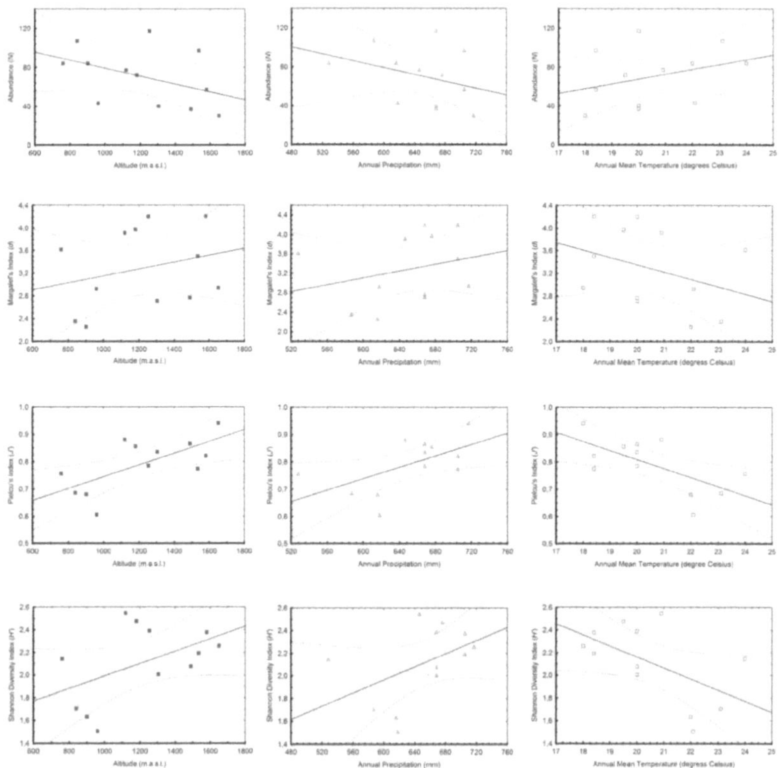